Praise for
Growing Conifers

One of the most misunderstood aspects of garden design is the proper use of conifers. They provide strong bones to any garden and *Growing Conifers* provides the key information needed to select the best plants for your garden. This book not only provides fantastic photographs to help you visualize how they will fit into your space, but it also provides all the information you need to properly plant and care for the new stars in your garden. After reading this book you may just decide to create a garden dedicated to only conifers.

—Robert Pavlis, author, *Soil Science for Gardeners* and *Building Natural Ponds*

In a world often devoid of beauty and landscapes lacking imagination, tools to correct the drab status-quo are always enthusiastically welcome. David's photography is artistry to the point of motivation and John's prose, based on decades of experience, is much more than sufficient to guide the home gardener to success, both in terms of plant selection and plant husbandry. *Growing Conifers* is a welcome addition to the bookshelf of any plant enthusiast and a must-add to the arsenal of those specializing in these captivating and collectable cone-bearing plants.

—David Olszyk, president, American Conifer Society

Growing Conifers is a work of love for the subject matter, reflecting a lifetime of devotion and a deep appreciation for the role of evergreens in our natural world. I am tempted to say that it's all here – that you will never need another book on this subject. It answers all questions including many that the reader would not think to ask. You owe it to yourself to wade into *Growing Conifers* with purpose and with time on your hands. It is thorough and passionately written and gorgeously illustrated. I will read it again and again and make it a centerpiece of my modern gardening library. I will turn the pages, slowly, in search of inspiration for my own garden. Like the best cup of coffee or finest wine you have ever tasted, *Growing Conifers* is not to be rushed. That would be a waste of a most incredible book.

—Mark Cullen, garden columnist, *Toronto Star*, member, Order of Canada, author, *The New Canadian Garden*

This is a wonderful, fantastic, encyclopedic book, worth owning for the information presented alone. Oh, but Albers writes like a poet and OMG, Perry's are not photos, but stunning photography! What an unusual and extremely winning combo.

Jeff Lowenfels, author, *DIY Autoflowering Cannabis*

Growing Conifers is a jewel of a book. Dr. John Albers invites the reader to take responsibility for greening and enriching the urban landscape by planting conifers. He includes everything a gardener needs to know, from planning a garden and choosing the right conifers, to planting, care and maintenance, and propagation. The book has depth, the photos are spectacular, the facts are scientifically accurate, and the charts help the reader find the right conifers for their area of the country. This is a must-have for any gardener's bookshelf.

—Brent Marcus, owner, Conifer Kingdom

This might be the most practical and handy book I've ever read on gardening with conifers. Being from the eastern Great Plains, I've always felt like we have few options, especially when it comes to sustainability, resilience, and wildlife habitat. *Growing Conifers* is written so gracefully and matter-of-factly that I feel empowered to try my hand at this almost mythical group of plants.

—Benjamin Vogt, prairie-inspired garden design, author, *A New Garden Ethic*

Growing Conifers contains an unbelievable amount of information for the reader. John goes into all parts of cultivation about conifers and their use in the landscape. It is an invaluable resource for the novice while also containing a wealth of information for the expert. It cannot go into detail on any one topic, yet it provides a firm foundation for anyone wanting to explore any one aspect of the conifer world in greater depth. It should be a part of the library of anyone working with conifers.

—Robert Fincham,co-founder, the American Conifer Society,
author of five books on growing conifers

This well-considered introduction to conifers as an important element of our ecosystems and gardens is brought to life with superb photography illustrating every key element of the text, from beautiful images of conifers in their native habitats and the taxonomic identification of cones, needles, and foliage, to garden plants, and design considerations. The author shares his 30 years of gardening with, and caring for, conifers as he learned to relish the shapes, textures, colors, and design possibilities that should allow conifers to be considered as the backbone of climate adapted gardens of the future.

—Saxon Holt, Saxon Holt photography, co-author,
Plants and Landscapes for Summer-Dry Climates

A rare and unusual combination of a superb gardener as author and a renowned photographer make *Growing Conifers* a one-of-a-kind book. The detailed descriptions of the many kinds of conifers, along with the excellent advice from John Albers, a long-term gardener, is superb. John has grown and cultivated most of these plants for years. The excellent photography of David Perry adds much clarity to the publication. It is a beautifully written and illustrated publication.

—John Wott, Director Emeritus, UW Botanic Gardens

GROWING CONIFERS

THE COMPLETE ILLUSTRATED GARDENING AND LANDSCAPING GUIDE

JOHN J. ALBERS

WITH PHOTOGRAPHY BY

DAVID E. PERRY

new society
PUBLISHERS

Cover design by Diane McIntosh.
Cover photograph © David E. Perry

All photographs © David E. Perry, unless otherwise noted.

Printed in Canada. First printing May 2021.

Any other inquiries can be directed by mail to

New Society Publishers
P.O. Box 189, Gabriola Island, BC V0R 1X0, Canada
(250) 247-9737

LIBRARY AND ARCHIVES CANADA CATALOGUING IN PUBLICATION

Title: Growing conifers : the complete illustrated gardening and landscaping guide / John J. Albers; with photography by David E. Perry.

Names: Albers, John J., author. | Perry, David E., photographer.

Description: Includes bibliographical references and index.

Identifiers: Canadiana (print) 20200415603 | Canadiana (ebook) 20200415697 | ISBN 9780865719569
(softcover) | ISBN 9781550927498 (PDF) | ISBN 9781771423458 (EPUB)

Subjects: LCSH: Ornamental conifers. | LCSH: Conifers.

Classification: LCC SB428 .A43 2021 | DDC 635.9/35—dc23

Funded by the Government of Canada | Financé par le gouvernement du Canada | Canada

New Society Publishers' mission is to publish books that contribute in fundamental ways to building an ecologically sustainable and just society, and to do so with the least possible impact on the environment, in a manner that models this vision.

MIX Paper from responsible sources FSC® C016245 | FSC www.fsc.org

Certified B Corporation

new society PUBLISHERS

Contents

I dedicate this book to my loving wife Santica

Acknowledgments

I wish to thank Sam Pratt for his review of an early draft of the book and particularly for his suggestions of conifer cultivars for inclusion in the book and the table on North America regions where conifers species can be grown. Special thanks to Dr. Ross Bayton for his thorough review of "Conifer Families," and to UW Professor Tom Hinckley and OSU Professor John Bailey for their review of "Conifer Basics." I am grateful to OSU Extension Forester Glenn Ahrens and UW Professors Robert Gara and Robert Edmonds for their review of conifer pests and diseases and to Paula Dinius, WSU Extension, for her helpful comments on "Designing with Conifers," "Planting Conifers," "Propagating Conifers," and "Sustainable Caring for Conifers"; to Ron Amos for his review of the chapter on propagation; and to Mona Gilkerson for preparation of the figure on North American regions. I wish to thank UW Professor Bruce Lippke and Dr. Elaine Oneil for their help on the section on carbon sequestration of conifers, and Anna Katz and Kristi Hein for editorial review of the early versions of this book. I am especially grateful to Cindy Riskin for her insightful suggestions for improving the final draft of the book. I am thankful to David E. Perry for providing the outstanding photographs. I really appreciate the efforts of the team at New Society Publishers for helping to build a more ecologically sustainable society through their books. Special thanks to copy editor Linda Glass, for her probing questions and excellent suggestions to further clarify the text, and to book designer Mary Jane Jessen for creatively making the book particularly visually attractive.

Introduction

Gardening has been my passion for more than three decades. I enjoy planting, watching the plants grow, and monitoring the changing plant communities through the seasons and years. Over the past 20 years, I have created and maintained a four-acre garden in Bremerton, Washington. I joyfully open this botanical oasis to the public for tours and for workshops on a variety of horticulture topics. Not only is Albers Marcovina Vista Gardens (www.albersvistagardens.org) a sanctuary for plants, wildlife, and visitors, it is also a laboratory in which I experiment with sustainable gardening best practices on a constantly expanding collection of plants. From this labor of love, I developed my first book, *Gardening for Sustainability: Albers Vista Gardens of Kitsap* (Vista Gardens Press, 2013), and my second book, *The Northwest Garden Manifesto: Create, Restore, and Maintain a Sustainable Yard* (Mountaineers Books, 2018).

Author John J. Albers in his garden.

Albers Vista Gardens showcases many kinds of plants, including 130 conifer species and over 400 conifer cultivars. After having cared for them for so many years, I've come to appreciate just how aesthetically, functionally, and ecologically versatile they are. My conifers will be a large part of my legacy, as they will continue to thrive long after I am no longer here. I hope this book will help others to select, grow, and care for conifers. When you plant a conifer, you are contributing to the health of your family, your community, yourself, and the environment.

I am fascinated by conifers. They are some of the oldest organisms on earth. The Great Basin bristlecone pine (*Pinus longaeva*), for example, can live for 5,000 years. Conifers are also some of the tallest, largest organisms on earth. California's magnificent giant redwoods (*Sequoia sempervirens*) can exceed 300 feet and have inspired feelings of spiritual transcendence for millennia. You can experience some of this awe without even leaving the city by simply planting conifers in your urban back yard. Conifers provide a sense of permanence and a feeling of well-being. They add a wonderful dimension to urban landscapes because their forms, textures, and colors contrast with deciduous plants and broadleaf evergreens. Because they vary in size, color, and form, they fulfill many

Ancient bristlecone pine in Great Basin National Park.

David Perry poses for his own remotely-triggered camera in the coast redwoods (*Sequoia sempervirens*) of Jedediah Smith Redwoods State Park.

Conifers have an innate ability to make other plants in their proximity stand out and show up even better. Notice the contrasting forms, colors, and textures among the conifers and the deciduous plants in this image.

roles in the landscape. You can plant conifers in containers, use them as visual anchors in mixed borders, employ them as groundcovers, or make them focal points. Midsize or large conifers can shade hardscapes, frame a vista, screen an unwanted view, back a flower bed or shrub border, function as a windbreak, or buffer sound. In spring, the colors of new conifer growth are especially bright and rich. In summer, their colors soften, letting flowers take center stage. In autumn, their greens and blues glow among the oranges, yellows, and browns of deciduous leaves. They brighten the dark days of winter, and their refreshing, lemony scents are delicious, no matter the season.

Some people think conifers are only firs or Christmas trees, but this vast and varied division contains more than 600 species. Conifers are woody trees, shrubs, and

groundcovers that produce naked (exposed) seeds, usually borne on cones. They are the largest group of nonflowering plants, or *gymnosperms.* Fossil records indicate that these resilient plants evolved more than 300 million years ago, before the Mesozoic Era of the dinosaurs. Much later—around 130 million years ago, in the Cretaceous Period—flowering plants, or *angiosperms,* evolved to have seeds enclosed within fruits.

Despite competition from thousands of different flowering plants, and despite climatic and geologic upheavals, conifers thrive in many regions of the world. Adapted to a wide range of sometimes harsh environments, conifers can grow in the bitter cold and deep snow of high mountains or on a rock wall in an arid region. They can deal with torrential rains and long summer droughts, blasting winds, and waterlogged soil. Because of their wide ecological versatility and their individual adaptability, conifers are suited to almost every garden in many temperate regions of the world.

Atlas cedar (foreground, right), 'Green Arrow' Nootka cypress (far left), and immediately behind it a glowing, golden-larch create a visual feast of form and color that amplify the rich, autumn reds of 'Omato' Japanese maple (center) in the author's garden.

Population growth, rapid urban development, increasing ambient carbon dioxide, climate change, and the loss of ecosystem services to the urban environment result in *urban heat islands.*

Urban areas tend to be 5°F to 10°F (3°C to 6°C) warmer than surrounding, less-developed areas. On warm summer evenings, this difference can reach 20°F. Fossil-fuel emissions, greenhouse gases, air pollutants, the absorption and release of solar radiation by dark-colored roofing and impervious paving, and high energy usage create this excess heat. A lack of trees, which filter the air and shade the ground, exacerbates this problem. Compared to rural areas, cities also have poorer air and water quality, greater noise and light pollution, and higher incidences of health conditions such as asthma and other respiratory illnesses, some kinds of cancer, and ischemic heart disease. Urban gardens help us adapt to and mitigate climate change and the urban heat island effect by providing shade, trapping air pollutants, and significantly cooling the air when they transpire (lose water) through their foliage.

Although adding any kind of flora to your yard—or even your apartment balcony—improves urban environmental quality, adding conifers is particularly effective. Because of their high leaf area index (leaf density) and evergreen foliage, they intercept rainfall and remove pollutants and particulates year-round.

According to a 2010 article, "Influence of Seasonality and Vegetation Type on Suburban Microclimate," outdoor spaces with a high leaf area index have lower soil, surface, and ambient air temperatures. They also tend to have better air and water quality.

Note that both native conifers and site-appropriate nonnative conifers provide similar ecological benefits. These ecological services are beneficial natural processes of healthy ecosystems, such as climate regulation, purification of air and water, and detoxification of environmental pollutants, which serve all

Rush-hour traffic and idling cars create a textbook example of an urban heat island and pollution.

Wind River near its outflow into the Columbia River Gorge. Conifers intercept rain and prevent pollutants from entering the watershed.

living organisms. However, both conifers and broadleaf plants are key components of the urban forest, providing overlapping and different benefits throughout the year. The following list describes the many ways conifers and broadleaf plants, particularly trees, can improve the urban environment.

- **Provide shade** by blocking solar radiation from the ground, pavement, and buildings, reducing the energy absorbed and reradiated back into the air, thereby lowering the temperature of the ambient air, soils, and impervious surfaces. Not only does this directly cool the area, but it also reduces the need for air conditioners and fans (and the factories required to produce them and the delivery trucks to transport them, etc.). This, in turn, reduces ozone-forming chemicals that trap other harmful compounds in the atmosphere in a well-known cycle that raises temperatures and impairs air quality.

- **Cool the air** through transpiration (evaporation through the leaves). In transpiration, plants draw water from the soil, pull it through water-conducting tissues to the leaves, and release it as vapor through *stomata*, tiny pores on the undersides of leaves. Transpiration consumes energy, converting liquid water to water vapor, so it cools the surrounding air. Because deciduous trees have a higher transpiration rate than conifers, they are, in general, better (when they have their leaves) at evaporative cooling than conifers.
- **Take up carbon dioxide** during photosynthesis, converting it into the oxygen we need to breathe and into sugars, which they store in plant tissue, making **conifers** the **largest terrestrial sink for carbon storage** (*sequestration*). Evergreen conifers photosynthesize year-round, removing more carbon dioxide annually than deciduous trees, which lose their leaves annually. Slow-growing, long-lived trees such as conifers can store more carbon over time than other trees. For example, a hectare (about 2.5 acres) of Douglas-fir in the Pacific Northwest can sequester as much as 350 tons of carbon in 100 years, and a hectare of Norway spruce in the mountains of Central Europe has been reported to store about 230 tons of carbon in 120 years. These old forests reach a maximum carrying capacity for their respective regions, and any additional carbon stored is offset by losses to mortality and decay. Thus, leaving a forest to grow to these advanced ages provides a limited one-time increase in carbon storage. In contrast, if the old-growth trees are used in construction, such as building houses and other structures, or used to make wood products, the trapped carbon in the wood remains sequestered, while the area of the old-growth forest can be planted with new trees that can continue storing carbon until they reach maturity. Sustainably managed Douglas-fir forests in the Pacific Northwest, which can be harvested and replanted with conifers on a sustained rotation of about 45 years, transfer carbon from the forest to the sustainable production of wood products as they are being used, recycled, or ultimately transferred to landfills. This process results in a sustainable increase in carbon stores year after year. These managed forests can displace and store about 3 tons of carbon/hectare/year indefinitely. Recycling a portion of the wood can further increase the amount of carbon stored. For example, if 40% of the wood is recycled, over 4 tons of carbon/hectare/year can be displaced and stored. Substitution of wood products for products requiring high levels of fossil fuels in their manufacture, such as steel and concrete, can further reduce greenhouse gas emissions. Sustainably managing conifer forests and effective use of their products can produce a combination of carbon sequestration and carbon emissions reduction and thereby play an important role in mitigating global warming.

- **Absorb other air pollutants,** such as carbon monoxide, sulfur dioxide, ozone, and nitrogen oxides, as well as particulate matter (small airborne particles less than 2.5 micrometers) that foul our lungs and pollute our skies. Because of stomata, grooves, and other foliar structures on conifers' needle-like or scale-like leaves, they are more efficient than broadleaf plants at capturing and filtering small particulates. Their year-round dense leaves also make conifers more effective at removing air pollutants, including ozone, both in winter and during hot, dry periods, when ground-level ozone levels are high. Their bark also intercepts and adsorbs some gases; this is particularly important under very dry conditions, which compromise stomatal activity. Deciduous trees, when in leaf, also remove pollutants.
- **Provide a windbreak**—a physical barrier that slows wind, even in winter. Especially important in winter in colder climates, windbreaks moderate cold winds, reducing cold air infiltration and heat loss from buildings and the consequent need for heating. In addition to cooling buildings in summer, conifers reduce energy usage year-round.
- **Intercept rainfall and absorb water-based pollutants,** which degrade water quality and aquatic habitat. Because it is so dense, conifer foliage intercepts much more rainfall than deciduous trees, which are ineffective at slowing stormwater in fall and winter. When rainfall runs off impervious roofs, roads, and sidewalks, it washes pollutants such as petroleum hydrocarbons, heavy metals, phosphorus and nitrogen from fertilizers, detergents, animal feces, and pathogens into our waterways. Although urban pollutants eventually reach waterways, slowing runoff allows more water-based pollutants to be absorbed by soil and plants. Moderating storm flows also reduces human health hazards from water pollution and flooding, such as waterborne illnesses. This effect is particularly important in regions like the Northwest, which typically experience most storms in fall and winter.
- **Minimize erosion** by anchoring soil with their roots, particularly on slopes.
- **Break up soil compaction** with their roots, lessening flood danger and dust pollution and enhancing water infiltration, rainfall storage, and water availability to other plants. By slowing water infiltration, roots filter pollutants as well.
- **Provide wildlife habitat,** particularly for the birds that enliven our lives, serenade our yards, and help balance insect populations. A yard with some well-chosen conifers—especially natives or site-adapted nonnatives—alongside a diversity of other trees, shrubs, and groundcovers offers food, shelter, and nesting sites to city-dwelling birds, mammals, amphibians, reptiles, and many types of pollinators. Conifer needles and twigs are particularly

Steller's jays preening and courting in a backyard weeping juniper.

important in winter, when other food sources are scarce. Many wildlife species, including songbirds, squirrels, and mice, eat pine seeds. In some areas, pine seeds also feed quail, chipmunks, voles, grouse, wild turkey, deer, bear, and hundreds of species of moth and butterfly larvae. The latter two, along with other needle-eating insects, attract many species of migrating and nesting birds. Then come the predatory birds, such as eagles, hawks, and owls, to prey on those smaller animals. Such is the cycle of life outdoors.

Conifers' dense, prickly foliage helps conceal nests and deter predators. And the shape of many conifers funnels water to the center of the plant, where birds and other wildlife can sip in safety. Pine siskins, juncos, hawks, golden-crowned kinglets, and warbler species are some of the many birds that nest in the dense foliage of hemlocks. Cavity-nesting birds, such as nuthatches, chickadees, owls, and woodpeckers, nest in the cavities they excavate from the soft wood of dead conifers, or they reuse old cavities rather than nest in hardwood trees.

Long-eared owl in a Douglas fir being harassed by a typically territorial pair of crows.

- **Provide a food source for people,** as the seeds of many conifers are edible. Most of these edible "pine nuts" are not sold as commercial crops because the size of the seed is too small, difficult to access, or birds and other wildlife eat the seed before it can be harvested. Yet, quite a few species of pine nuts are harvested to provide a local food source. The main source of pine nuts from North America is the piñón pine, chiefly the larger pine nut from the single-leaf piñón (*Pinus monophylla*), native to the American Southwest, and the smaller, milder pine nut from the two-needle piñón (*Pinus edulis*), native to the American West, along with the Mexican piñón (*Pinus cembroides*).

Two Asian pine species, the Korean pine (*Pinus koraiensis*) and chilgoza pine (*Pinus gerardiana*), are widely harvested and exported. The main pine nuts from Europe come from the stone pine (*Pinus pinea*), which have been cultivated for thousands of years. Roasted pine nuts eaten in large quantities by the Australian native peoples are from the Bunya-pine (*Araucaria bidwillii*), while the large seeds of the Japanese nutmeg-yew (*Torreya nucifera*) are sold as a dessert or pressed to make vegetable oil. The inner bark and sap of members of the pine family have high levels of vitamin A and C and many

Piñón pine forest, *Pinus monophylla,* in Great Basin National Park. Inset: Piñón seeds are a valuable food source for humans and a valuable commercial crop. Piñón-juniper ecosystems have had subsistence, cultural, spiritual, economic, aesthetic, and medicinal value to Native American peoples for centuries, and single-leaf piñón has provided food, fuel, medicine and shelter to Native Americans for thousands of years.

other nutrients. Strips of the inner bark can be cooked and eaten or dried and ground into a flour for making bread or adding to soups. Years ago, American explorers used conifer bark to treat scurvy. The sap of pine trees can be tapped in spring and drunk as a tea, and young pine needles rich in vitamin C can be steeped to make tea. Protein and vitamin-rich pine pollen can be used as a flour additive and cooked with a large variety of foods.

- **Visually buffer unwanted views,** such as your neighbor's garbage bins or the trash-filled lot across the road.
- **Provide privacy and security** by deterring potential thieves looking for vulnerable houses.
- **Benefit mental and physical health** by providing year-round *green infrastructure* in our yards and our communities. Green infrastructure is an interconnected network of natural and seminatural areas designed to protect and restore natural processes. Urban landscapes are essential for outdoor physical activity and social gatherings, which are causally associated with good mental well-being, cognitive function, and the reduction of chronic stress. Evergreen conifers are especially important for providing these benefits in winter and therefore are an essential part of green infrastructure.

Snow on the variegated needles of the Japanese white pine *Pinus parviflora* 'Ogon'.

Conifers, in general, are more resistant to water stress than broadleaf trees because of several anatomical and physiological adaptations. Like broadleaf woody plants, conifers store water in their trunks, stems, and leaves. Unlike broadleaf trees, however, they have large water storage reservoirs in their sapwood and foliage, and their water-conducting cells refill more readily. Additionally, unlike in broadleaf trees, conifers' water transport system, or *xylem*, exclusively comprises *tracheids* (very small columns of long, narrow cells) that can sustain high tension in the water column and minimize the formation of air bubbles (*cavitation*), which can block water transport. Under extreme water stress, tracheids may allow a small air bubble to form, but they can close off and compartmentalize it, allowing the water transportation to continue.

Conifer leaves have several advantages over deciduous leaves when water is in short supply. Because conifer leaves

Bald eagle perched on a Douglas-fir in an old, established, Seattle neighborhood.

are smaller and narrower than in broadleaf plants, they lose heat more readily. This lowers their respiration and transpiration rates, so they lose less water during drought. A waxy *cuticle* further minimizes water loss by covering the epidermal cells on the leaf surfaces. The stomatal pores are sunken, and they have guard cells that open and close to permit or stop gas exchange. These conserve water during photosynthesis by tightly closing during hot, dry conditions or in freezing temperatures, when water is limited or when photosynthesis is not taking place. As deciduous trees have considerably higher transpiration rates than conifers, they lose more water to evaporation.

It is for all these reasons that we need more conifers in our cities and towns, and, therefore, I wrote this book. Given the horticultural and ecological importance of urban conifers, it is vital that all of us do our part to restore conifers to our urban environment.

Conifer Basics

Chapter 1

From the wind-twisted spruce that tops a rocky embankment to the graceful Douglas-fir that shelters an understory of ferns to the dwarf conifer in the large vase on the patio, there's a conifer to suit almost any setting and purpose. But the choices can be overwhelming. It's well worth taking time to learn the basics of taxonomy, plant features, and specific characteristics before shopping for conifers so you will be able to select the best conifers for your garden.

The Linnaean system consists of a hierarchy of groupings or *taxa*. From largest to smallest groups, the plant taxa are

kingdom → division → class → order → family → genus → species

- A *family* is a group of related *genera* (the plural of *genus*).
- A *genus* is a group of related *species*.
- A *species* is a natural population or group of populations whose form is distinct, and whose individuals can interbreed to produce fertile offspring. A single species is abbreviated "sp." and multiple species "spp."
- A *subspecies* (abbreviated "subsp." or "ssp.") occurs in some species, particularly those distributed widely. Plants earn a subspecies designation when slight but consistent differences are evident in color, height, or form.
- A *variety* (abbreviated "var.") has a consistent trait slightly different from the species. Varieties often occur and reproduce in nature, and most varieties are *true to type*—that is, seedlings will have the same unique traits as the parent plant. For example, *Pinus contorta* has three naturally occurring varieties (sometimes referred to as subspecies): *Pinus contorta* var. *contorta,* the shore pine; *P. contorta* var. *latifolia,* the lodgepole pine; and *P. contorta* var. *murrayana,* the Sierra lodgepole pine. (Note: When the genus is repeated in writing, it may be abbreviated as shown.)

Taxonomy

Taxonomy is the science of plant classification founded by Swedish physician Carolus Linnaeus. It groups plants based on similar key features, which explain how groups of species evolved. Originally, classification was based solely on physical attributes. Since the advent of molecular biology, scientists can inspect an organism's very DNA to classify it.

Cultivar Terms

A cultivated variety, or cultivar (abbreviated "cv."), has at least one attribute, such as leaf color or shape, that distinguishes it from the species and makes it worthy of propagation by plant nurseries. Although some cultivars originate from mutations and others from hybridization of two different plants, a cultivar is selected and cultivated by people. A cultivar is not necessarily true to type—propagating by seed usually produces a plant different from the parents—and often it must be propagated vegetatively, such as by cuttings or grafting.

Understanding these terms will be helpful as you read the text.

- **Provenance:** A local population of a native species that contains genetic variations adapting it to the local environmental conditions.
- **Clone:** An individual that is genetically identical to another individual.

- **Seedling selection:** A chance seedling with unusual desirable traits that someone selects and then breeds for those traits.
- **Witch's broom selection:** A conifer grown from cuttings taken from a *witch's broom*—a tightly congested formation of twigs and foliage due to a genetic mutation that has the characteristics of that witch's broom.
- **Dwarf seedling:** A chance seedling that grows significantly slower than the original plant.
- **Sport:** A stable mutation in foliage color or texture. Many variegated and golden varieties of conifers are derived from sports.

Classifying Conifers

Most plants are given a common name in the language where they are native, and a scientific name that is Latinized. For most people, common names are easier to remember and pronounce, but using common names can lead to confusion or be misleading. Common names contain two or more words; the last word generally denotes the genus or plant group the plant belongs to, and the first word(s) describe the plant. In some cases, however, the common name does not denote the proper plant group, such as the conifers western red-cedar or Douglas-fir. The western red-cedar is not a true cedar, and the Douglas-fir is not a true fir. To alert the reader to those instances where the conifer's common name does not denote the correct genus, a hyphen has been added between the words of the common name.

Plants often have more than one name, even in a single region, and common names, such as *cedar,* are applied to many genera other than *Cedrus,* the true genus for cedars. To avoid this problem, botanists unambiguously give each plant a *binomial* (two-term) name: the genus followed by the species. The genus and species are always italicized; the genus is capitalized, and the species is lowercase. Garden conifers often have a cultivar name, which is not italicized but is capitalized and enclosed in single quotation marks. For example, in the binomial *Pinus contorta* var. *latifolia* 'Chief Joseph', the genus is *Pinus,* the species is *contorta,* and 'Chief Joseph' is the cultivar of the tree commonly known as as lodgepole pine 'Chief Joseph'. Compare how different the cultivar 'Chief Joseph' is from the common lodgepole pine. The common lodgepole has year-round dark-green foliage, and the 'Chief Joseph' has green foliage that becomes brilliant golden in autumn and winter.

Botanists often change conifer classification, and sometimes their names, as new knowledge improves our understanding of conifer genetic relationships and evolution. For this book's purposes, we will use the current classifications and the more-obvious physical traits that you can use to identify conifers in your yard and community. The following section gives you the main identifying characteristics to use in identifying conifers.

Leaves

- **Shape**
  - Narrow needle-shaped leaves, as in the pines (genus *Pinus*).
  - Flattened blades, as in redwood (*Sequoia*), yew (*Taxus*), and hemlock (*Tsuga*).
  - Flat scale-like leaves, as in the genera false-cypress (*Chamaecyparis*), arborvitae (*Thuja*), and true cypress (*Cupressus*).
  - Awl-shaped leaves tapering to a slender, stiff point, as in the juvenile foliage of many junipers (*Juniperus*).
  - Flat strap-like leaves, as in plum-pines (*Podocarpus*), which also have other common names.
- **Cross-section:** For example, spruce (*Picea*) needles have a quadrangular cross-section, and firs have a flat cross-section.

Flat, scale-like leaves in false-cypress (*Chamaecyparis obtusa*) (left), and in arborvitae (*Thuja plicata*) (right).

Flat, strap-like leaves of plum-pine *Podocarpus alpinus* 'Blue Gem'.

- **Attachment:** Where the leaves attach to stems, twigs, and *branchlets* (small branches or divisions of small branches, especially terminal divisions; the term is usually applied to branches of current or preceding year):
  - Needles fastened in tight bundles at the base by a sheath, or *fascicle,* as in pines.
  - Needle clusters radiating out from a peg-like bud without a basal sheath, as on the short shoots of larch (*Larix*) and true cedar (*Cedrus*).

Needle clusters of the true cedar *Cedrus deodara* 'Snow Sprite'.

Needles attached directly to the stem in the Min fir *Abies recurvata* var. *ernestii.*

The spiral needles of Douglas-fir are twisted at their base and therefore appear two-ranked.

- Single needles spread along the branch from a peg-like stalk, as in spruce (*Picea*).
- Single needles attached directly to the twigs and branchlets, as in true fir (*Abies*), Douglas-fir (*Pseudotsuga menziesii*), and hemlock (*Tsuga*).

- **Arrangement:**
  - Opposite, as in dawn redwood (*Metasequoia glyptostroboides*).
  - Alternate, as in bald-cypress (*Taxodium distichum*).
  - Spiral, as in true firs and Douglas-fir, but twisted at the base to appear two-ranked.
  - Whorled, as in juniper.
  - Spiral, as in Japanese-cedar (*Cryptomeria*) and giant sequoia (*Sequoiadendron*).
- **Number in a cluster:**
  - Two, three, or five needles per cluster, as in pine.
  - Greater than five, as in larch or cedar.

Pine needles have two, three, or five in a cluster.

- **Markings:** White bands caused by a waxy coating, or white bloom, around the tiny stomata.
  - Two white stomata bands on the lower surface, as in grand fir (*Abies grandis*), Pacific silver fir, and western hemlock.
  - Two stomata bands evident on the lower surface and one on the upper surface, as in subalpine fir (*Abies lasiocarpa*).
  - Two stomata bands evident on the upper and lower surface, as in noble fir.
  - No distinct stomata bands, as in yew.
  - On lower surface, in the shape of a butterfly, as in western red-cedar.
  - On lower surface, in the shape of an "x," as in Port Orford-cedar.

White stomatal bloom on the bottom of the leaves (left to right) western red-cedar, grand fir, and dwarf Hiba arborvitae.

- **Timespan:** Deciduous conifers, such as species of larch at high altitudes or in subalpine regions, retain their leaves for as little as five months. Other deciduous conifers, including *Taxodium*, *Metasequoia*, and *Pseudolarix*, retain their leaves from spring to fall. Most evergreen conifers retain their leaves for two to six years before dropping them. However, some evergreen conifers are exceptions to this rule. The Pacific silver fir retains its foliage for up to 22 years, and the Great Basin bristlecone pine (*Pinus longaeva*) and some *Araucaria* species can retain their leaves for up to 50 years. Because evergreen conifers lose a small proportion of their leaves annually, they don't need to produce a complete set of leaves each year, helping them succeed on nutrient-limited soils. The oldest interior needles on most evergreen conifers—such as those of pine, spruce, fir, and arborvitae—turn yellow or brown and drop off in the fall, while the remaining leaves at the tips of the branches stay green. The change in needle color and needle drop is a normal physiological response to cooler nights and shorter days. In contrast, some needles on evergreen yews turn yellow and drop in late spring and early summer, when the temperature warms. Old interior needles can also shed because of damage by pests, such as spider mites, which cause yellowish spots or stippling; pathogenic fungi, such as Rhizosphaera needle cast disease; or environmental factors, such as drought and nutrient deficiency.
- **Fragrance:** Many conifers have a distinctive scent when the foliage is crushed.

Branching

- Conifers with whorled branches that form a circular pattern around the central leader or central portion of the tree, such as pines, spruces, true firs, true cedars, and Douglas-firs, have one growth spurt each spring at branch tips and the top of the central leader, from buds formed the previous year. These conifers don't have latent buds on old wood.
- Conifers with a random branching habit, including arborvitae, false-cypress, yew, juniper, and hemlock, have multiple growth spurts throughout the growing season and have many latent buds along their stems.

Cone Position

Some conifers, such as yews and many junipers, have male and female cones on different plants. These conifers are *dioecious* (two houses). However, most conifers, such as pines, bear both on the same plant. These are *monoecious* (single house). Male and female cones

Seed cones in the upper crown and pollen cones below the seed cones in the Korean fir (*Abies koreana*). (This photo was taken at the beginning of May.)

of monoecious conifers usually grow on different parts of the conifer and/or become fertile at different times, minimizing self-pollination. For example, pollen cones (male) are often in the mid- to lower crown of the fir, whereas seed cones (female) are in the upper crown.

Pollen is unlikely to float up to the female cones above. This arrangement increases genetic diversity by minimizing self-pollination, as cross-pollination between two plants enhances the species' adaptability to a changing environment. Also, the higher the seed cones are up in the tree, the farther the fertilized seed cones are likely to disperse.

Conifer seed cones come in a stunning variety of sizes, shapes, textures, and colors, providing year-round visual appeal.

A

B

The Korean fir's cones change with the seasons. (Photo A taken in late June, photo B captured in late October, photo C captured in early February, photo D captured in late February after the freezes and thaws of winter, and photo E shows rachises photographed in early May.) All these images except C were taken from the same tree in the author's garden.

Seed Cones

Seed cones are woody and come in a large array of sizes, shapes, and colors. The following traits can help you identify conifers:

- **Shape:** Spherical, egg-shaped, oblong, conical, cylinder, or squashed sphere.
- **Color:** Can change dramatically with the seasons, with a few species exhibiting bright, showy colors in spring. For example, the cones of Korean fir, *Abies koreana*, goes through colorful stages, which is one of the reasons people love to have this fir in their garden. The seed cones emerge purple around May, transform to a bright blue in June, and gradually turn brown in autumn. By the following February, the cones are brown and resinous, and by spring, the old cones fall apart, leaving behind the central spikey rachis, which eventually disintegrates.

D

C

E

- **Attachment:** Seed cones can be upright on a branch, as in true fir, or pendent, as in spruce and hemlock. In addition, seed cones can be terminal, as in spruce and hemlock, or along the branches, as in true firs, most pines, and Douglas-fir.
- **Size:** From as little as ⅕ inch (1 cm) long, as in the small cones of some junipers and *Microbiota decussata,* to as large as 28 inches (70 cm) in the sugar pine (*Pinus lambertiana*).
- **Timespan:** Cones can persist for only one year, as in the white spruce (*Picea glauca*), or remain on the branch for several years, as in the black spruce (*Picea mariana*). Pines that retain unopened seed cones, such as the lodgepole and whitebark pine, originate in fire-prone ecosystems. These cones open only when exposed to fire.
- **Time of appearance:** Most seed cones first appear in the spring; cedar cones don't appear until late summer or fall.
- **Time of maturation:** In most conifers, seed cones mature one year after pollination. In some conifers, such as cedars, cypresses, and many pines, seed cones mature in two years. In a few pines, seed cones can take three years to mature.
- **Cone scale arrangement:** Arrangement can be *imbricate,* with cone scales spirally arranged and attached along a common axis, with edges of the cone scales overlapping, as in the pine family (Pinaceae), or *peltate,* with cone scales appearing to be attached at a central point and projecting out from that point like an umbrella, as in some members of the cypress family (Cupressaceae).

Cone scale arrangement varies widely in pine seed cones.

- **Cone scale variation in form and size:** For example, cone scales of five-needle pines have a terminal thorny outgrowth, called an *umbo*, in the middle, with no prickle. Three-needle pines have scales with a dorsal umbo and a yellow prickle.
- **Arrangement of bract scales with seed scales:** Bract scales are distinctive in some members of the pine family, such as Douglas-fir, and in many true firs and larches.

The bract scales in the Douglas-fir seed cone look like rat tails and hind limbs.

Scaly seed cones are common in most conifers that grow in temperate climates—the conifers that most North American gardeners grow. Conifers without scaly seed cones include juniper, podocarp (*Podocarpus*), yew, and plum-yew (*Cephalotaxus*).

As in other seed plants, the pollen-producing cones are male, and the seed-producing cones are female. The scales of the seed cones protect the developing seeds.

The seed cones of juniper are often referred to as juniper berries. They are an excellent nutritional food source for wildlife, such as the American robin, when other food sources are scarce in the depths of winter. Juniper berries are also valued by humans, who use them as the predominant flavor ingredient in gin.

Pollen Cones

Pollen cones are herbaceous, smaller, and shorter-lived than seed cones; they usually wither away once they release their pollen. However, pollen cones tend to be much more numerous than seed cones, and they can exhibit conspicuous colors.

Example of attractive pollen cones on *Pinus coulteri x arizonica* at Bloedel Reserve on Bainbridge Island, Washington (top) and on *Abies koreana* 'Horstmann's Silberlocke' (top, page 31).

Pollination and seed dispersal

Unlike flowering plants, which are primarily pollinated by insects and other animals, most conifers are pollinated by the wind. Wind-pollinated conifers usually bear both male (pollen cones) and female (seed cones) structures on the same plant (monoecious). However, some conifers, such as junipers, yews, and most large-seeded pines, need animals for seed dispersal. Animal-pollinated conifers usually bear either male or female structures, but not both.

Wind-dispersed seeds usually weigh less and are more numerous per conifer than animal-dispersed seeds. Lightweight seeds can be carried long distances by the wind. However, large seeds have more stored nutrients for their initial phase of growth before they have the capability to produce their own food through photosynthesis, giving them a competitive advantage over the smaller wind-dispersed seeds. To aid in dispersal, some conifers seeds have wings that allow them to move in a helicopter-like fashion away from the tree. Some wings are firmly attached, as in five-needle pines, and some wings are loosely attached, as in three-needle pines. Seed cones open when the weather is dry, and close when the weather is wet to ensure that the seeds will be well dispersed dry seeds fly much farther than wet seeds. Because wind pollination is inefficient, conifers need to produce and disperse billions of tiny pollen grains into the air, where they float and find their way to the ovule of a female cone for fertilization. However, only a lucky few of the pollen grains will encounter and enter an ovule of the same species and initiate fertilization. Broad seed dispersal helps conifers reach specific habitats that are favorable to survival and allows conifer species to migrate to new habitats, which is important for the species to adapt to a changing climate.

Buds

Buds are small new growths on the sides or ends of a branch in the form of protuberances on the stem that develop into a leaf or shoot. Some conifers, such as cypress, juniper, and arborvitae, lack distinguishable buds.

- **Shape:** Globose, ovate, round, cylindrical, spindle form, acute.
- **Bud scales:** These specialized protective leaves surround the buds of certain plants, either loosely covering or tightly appressed, at the apex straight or curved backward at the tip, with or without resin, and with hairs (e.g., the elongated buds sheathed in silky white hairs of *Pinus thunbergii*) or without hairs.

Twigs

- **Texture:** Smooth-surfaced without projections, as in firs, or rough-surfaced, with woody peg-like projections, as on spruce branches and the short shoots of true cedars and larches.

Bark

- **Texture:** Exfoliating in strips or in patches, scaly, furrowed, smooth or rough. In spruce, rough and scaly; in fir, smooth and furrowed.
- **Color:** Gray, brown, green, white, orange, or red—or a combination.
- **Thickness:** Conifer bark is generally thick, which provides protection from fire. For example, the bark of the giant sequoia (*Sequoiadendron giganteum*) can have bark more than 20 inches thick to protect it from fire.

Trunk Shape

- Swollen at the base, as in bald-cypress.
- Fluted and buttressed, as in dawn redwood.
- Nearly cylindrical with little taper, as in giant sequoia or Sitka spruce.
- Obvious taper, wider at the base and smaller at the crown, as in most other conifers.

The fluted and buttressed trunk of dawn redwood.

Well-fortified with all this botanical information, we will explore the conifer families and their geographical distribution and key features in the next chapter. Nearly all the temperate genera have the potential to be used in a sustainable garden.

Conifer Families

Chapter 2

The 614 conifer species currently found on earth represent 70 genera classified in the following eight families:

- Pine (Pinaceae)
- Cypress (Cupressaceae)
- Yew (Taxaceae)
- Plum-yew (Cephalotaxaceae)
- Umbrella-pine (Sciadopityaceae)
- Podocarp (Podocarpaceae)
- Celery-pine (Phyllocladaceae)
- Araucaria (Araucariaceae)

Let's look at each of these families in turn, along with the similarities and differences among the genera.

Family: Pine (Pinaceae)

The pine family Pinaceae is by far the most ecologically and economically important conifer family, providing most of the world's softwood timber used for the construction and paper-products industries. The members of this family, which includes the pine genera, not only provide shade, screening, and aromatic fragrance in the garden but are also excellent for wildlife habitat and soil erosion control. The family is dominant in the boreal forests of North America and the Pacific regions, and the genera and species in this family are found naturally in the three northern continents of America, Asia, and Europe, though they also occur farther south in Central America, the Caribbean, North Africa, and just one location south of the equator, Sumatra. They are primarily resinous and aromatic evergreen trees with narrow, needle-like leaves borne singly on long shoots or in tufts or tight bundles (fascicles) on short shoots in a spiral arrangement and leaves sometimes twisted. Major branching usually consists of multiple tiers of horizontal branches. The seed cones and pollen cones are usually found on the same plant (monoecious, or one home), but some conifers have them on different plants (dioecious, or two homes). The seed cones may be erect or dangling. Among the 12 genera in the pine family, the most important genera commonly seen in cultivation are as follows:

- Pine (*Pinus*)
- Spruce (*Picea*)
- True fir (*Abies*)
- Douglas-fir (*Pseudotsuga*)
- Hemlock (*Tsuga*)
- Cedar (*Cedrus*)
- Larch (*Larix*)

Pines

Pines (*Pinus*) are evergreen trees or shrubs distributed throughout the world but native primarily to the northern temperate regions; some grow naturally in the tropics. The trees' crowns are usually cone-shaped when young but often flat-topped or rounded with age. The leaves are needle-like, with two to five needles spirally arranged in fascicles or clustered on shoot spurs—except for the North American piñón pine (*Pinus monophylla*) which has a single needle in each fascicle. Pollen cones are small and numerous, arranged in dense clusters around the base of the current year's growth. The mostly woody seed cones have overlapping cone scales (imbricate), and they hang pendent and mature in two to three years. Cone form varies widely, from egg-shaped to elongated conical or nearly cylindrical, as does size, from less than an inch (2.54 cm) to up to 16 inches (40.64 cm). The seed cones take from six months to three years to mature, depending on the species. Some seed cones open at maturity and release their seed; others have *serotinous* cones; that is, they remain closed and sealed with resin until opened by fire.

Table Mountain pine (*Pinus pungens*) could brighten up almost any winter landscape.

Spruce

Spruces (*Picea*) are timber evergreen trees native to the northern temperate and boreal regions of the northern hemisphere. They have a spire-like or broadly conical crown, whorled branches, and thin scaly bark. Sharp-pointed needles are borne singly and spread radially from the twigs. The leaves are mostly four-sided and rigid, growing directly from the twigs without a stem or stalk, and are often forward pointing and upswept. They persist for up to ten years, and the woody peg-like base remains on the twig after the leaves fall. Pollen cones are single or grouped, shedding their pollen in the spring. Seed cones are usually pendent and borne primarily on upper branches.

Colorado spruce (*Picea pungens*) 'Straw' is aptly named and at its cream-tinged best in winter foliage (next page).

True Fir

True firs (*Abies*) are medium to large evergreen trees primarily found in cool, humid mountain ranges in the northern hemisphere. They have a typical spire- or cone-shaped crown, a single straight trunk, and regularly spaced horizontal branch whorls, with one whorl produced each year. The strikingly regular branch pattern is created by a single terminal with multiple lateral shoots forming each year at the tips of growing branches. The often-twisted, spirally arranged leaves have two prominent bands of stomatal bloom beneath, and they often have additional single lines of stomatal bloom on top. The leaves are borne singly and directly from the branch, persisting for five or more years. Where a leaf is detached, a flat, round leaf scar is left on the twig. Pendant pollen cones appear in dense clusters on the undersides of the current year's growth. In contrast, barrel-shaped or cylindrical seed cones are single and attached upright on stalks in the gap between a needle and the shoot beneath it in the upper third to upper quarter of the shoots at the top of the crown. Seed cones mature in a single season, and after a mature cone falls apart to release its seed, the spike-like central core of the cone (the *rachis*) remain on the branch.

Sicilian fir *Abies nebrodensis* 'Sicilian Gold'. This species is critically endangered, with only about 25 mature trees surviving in the wild (next page).

Douglas-fir

Douglas-firs (*Pseudotsuga*) are evergreen trees distributed in western North America and eastern Asia. The hyphen in the common name lets us know that Douglas-fir is not a true fir—that it's not a member of the *Abies* genus. *Pseudotsuga menziesii,* a native to the West Coast of North America, is an economically important tree species because of its extensive use in industry. Douglas-firs have a single straight trunk and numerous spirally arranged horizontal branches. Alternate, short-stalked, linear, flat leaves with a groove on top and two stomatal bands beneath are borne singly and either radiate in all directions from the branch or are set two-ranked on the branch. Small (about 1-inch-long [2.54 cm]), catkin-like pollen cones drop large quantities of yellow pollen in spring. Seed cones hang in a pendent arrangement like that of spruce, but what makes the Douglas-fir seed cone distinctive is its three-pointed bracts extending beyond the cone scales, with a long central lobe that looks like a mouse tail.

The nearly spherical pollen cones hang in the axils of leaves near the tips of the twigs.

Hemlock

Hemlocks (*Tsuga*) are distributed in eastern and western North America, eastern Asia, and the Himalayas. They are primarily cone-shaped evergreen trees with a leading shoot that usually droops, and their horizontal branches are usually arranged in flattened sprays that arch downward. The linear leaves have two stomata bands on the underside and, in most species, are spirally arranged and predominantly twisted into a flat row on each side of the twig (two-ranked). The leaves persist for several years. The sparse, nearly spherical pollen cones hang singly in the axils of leaves near the tips of the twigs. The seed cones, which hang at the tips of second-year twigs, mature in about six months; once they open to shed their seed, they fall intact after one year.

Atlas cedar *Cedrus atlantica* 'Glauca'.

True Cedar

True cedars (*Cedrus*) are distributed in northwest Africa, the Mediterranean region, southwestern Asia, and the western Himalayas. They are tall evergreen trees, ordinarily with a single straight trunk, a broad crown, and an erect or sometimes bent top. As cedars reach old age, the top of the tree becomes increasingly flattened. Spirally arranged solitary needles, three- or four-sided in cross-section, are borne on long terminal shoots; tufts of needles are borne on short shoots. Leaves persist for three to six years. Upright pollen and seed cones attach at the tips of short shoots. Seed cones mature in two to three years, then break apart to shed the seeds and seed scales.

Dahurian larch (*Larix gmelinii*) seed cones and needles.

Larch

Larches (*Larix*) are distributed across the boreal forest in the northern hemisphere and south of this region in the northeastern and northwestern United States and in the Himalayas and the Alps. They are deciduous trees with sparse, open, conical crowns and branches in a whorl with both long lateral shoots and short spur shoots. The long shoots sparsely bear well-separated, spirally arranged needles. The spur shoots are long-lived and bear many new tufts of needles (10 to 30 needles per spur) each year and often bear pollen cones and seed cones as well. Pollen and seed cones are borne upright at the tips of short shoots. The empty seed cones persist on the tree several years after seed release.

Japanese larch *Larix kaempferi* 'Paper Lanterns'. This photo illustrates that larch trees lose their needles in winter.

Family: Cypress (Cupressaceae)

Conifers of the cypress family were distinguished from the old Taxodiaceae family by their leaf arrangement, which is opposite or whorled instead of spiraled. When the dawn redwood (*Metasequoia glyptostroboides*) was first described in the late 1940s, it was found to be a close relative of two members of the Taxodiaceae, the coast redwood (*Sequoia sempervirens*) and the giant sequoiã (*Sequoiadendron giganteum*), even though it has an opposite leaf arrangement, like some members of the cypress family. Eventually, molecular studies provided evidence that the old Taxodiaceae family and Cupressaceae share a common ancestor, and taxonomists merged Taxodiaceae into Cupressaceae, except for the endemic koyamaki, or Japanese umbrella-pine (*Sciadopitys*).

The current cypress family, with its aromatic and resinous trees or shrubs, has 25 genera—the most of any conifer family. The best-known are

- False-cypress (*Chamaecyparis*)
- Japanese-cedar (*Cryptomeria*)
- True cypress (*Cupressus*)
- Bald-cypress (*Taxodium*)
- Juniper (*Juniperus*)
- Dawn redwood (*Metasequoia*)
- Giant sequoia (*Sequoiadendron*)
- Coast redwood (*Sequoia*)
- Incense-cedar (*Calocedrus*)
- Arborvitae (*Thuja*)

The cypress genera are all monoecious, except for *Juniperus*, which usually bears the seed cones and pollen cones on separate plants (dioecious). Cupressaceae is the most widely distributed gymnosperm family, occurring in diverse habitats and on all continents except Antarctica. However, among the 25 genera recognized by the Gymnosperm Database (the web's premier source of information on the classification, description, ecology, and use of conifers and other gymnosperms), 15 have a single species (monotypic), and many genera and species have a very restricted distribution. Most are evergreen, but three genera—*Taxodium*, *Glyptostrobus*, and *Metasequoia*—are deciduous. The leaves of many Cupressaceae genera are needle-like when young and become scale-like and pressed against the often-flattened branchlets as they mature. The leaves are arranged either spirally or in opposite pairs, with each pair 90° to the previous (opposite decussate), or in decussate whorls of three or four. However, the genera *Sequoia* and *Metasequoia* have needles arranged in two ranks. Awl-like juvenile leaves are often present on mature plants. In many species, the foliage spreads right or left of the twigs in a plane parallel to the ground, and the branches often turn up at the tips as the conifer matures. Pollen and seed cones are borne on the tips of short branchlets. The seed cone scales are arranged either whorled, opposite, or spirally.

Hinoki-cypress *Chamaecyparis obtusa* 'Nana Lutea,' a dense dwarf conifer with tiny, flattened fans of bright yellow foliage.

False-cypresses (*Chamaecyparis*) are native to Japan, Taiwan, and the east and west coasts of the United States. Several taxonomists now include the Chinese Fujian-cypress (*Fokienia hodginsii*) in *Chamaecyparis*, thus expanding the range of this genus. False-cypresses are pyramidal evergreen trees clothed in fibrous bark that peels in thin strips or becomes furrowed as the tree ages. Open-grown trees are densely branched from the base, with the branches spreading and arching upward but nodding at the tips, particularly those of the leading shoots. The branchlets form flattened or fan-shaped horizontal or drooping sprays. Young juvenile leaves end in a point but soon mature into overlapping, scale-like leaves in alternating pairs; the lateral pairs, but not the facial pairs, tend to wrap around the twig. Numerous pollen seed cones appear at the tips of short branchlets. Seed cones mature and open six to eight months after pollination.

Japanese-cedar *Cryptomeria japonica* 'Elegans' in winter.

Japanese-cedars (***Cryptomeria***) are native to Japan, one small region in central China, and one small region in southern China, although they are widely cultivated throughout China. They are evergreen trees with a straight trunk; the crown, conical when young, becomes cylindrical with age. This genus is quite variable, giving rise to the many cultivated varieties used in horticulture. The reddish-brown bark peels off in strips. Spirally arranged, needle-like leaves persist for four to five years. There are two or more stomatal bands on all four-leaf surfaces. Oblong pollen cones are clustered near the tips of the twigs but borne solitary. At the ends of down-curved branches, seed cones are borne solitary or in groups.

True cypresses (***Cupressus***) are ornamental and timber evergreen trees and shrubs distributed throughout the warmer temperate and subtropical regions of North America, Europe, Asia, and North Africa. Various organizations, like the California wild plant

Vietnamese cypress (*Cupressus vietnamensis*) exhibits both juvenile (right) and mature (left) leaf forms.

database Calflora, accept the splitting of *Cupressus*, with the old-world species remaining in that genus and new-world species designated as *Hesperocyparis*. The newest member of this genus, the golden Vietnamese cypress (*Cupressus vietnamensis*), is often placed in *Xanthocyparis*, as the taxonomy of this group is messy. This species was discovered on a high ridge in northern Vietnam in 1999. In most species, the bark separates into thin strips that shed from the plant. The leaves are awl-shaped and spreading on young plants, but on older branchlets they transition to small, scale-like leaves, pressed close to the branch. The leaves, with granular pits on the outer surface, cover the shoots in opposite groups, usually in opposite pairs, making the branch appear four-sided. Numerous spherical to oblong pollen cones are borne at the tips of short branchlets. The small, usually spherical seed cones, with three to six woody or leathery scales that have small projections on the outer surface, are borne at the tips of side branchlets. Seed cones mature at the end of the second year but may remain on the conifer for several more years before they open. One characteristic of *Cupressus vietnamensis* is its tendency to show both juvenile and mature leaf forms alongside each other.

Bald-cypress (*Taxodium*) is a genus of two species, *Taxodium distichum* and *Taxodium mucronatum*. They are important ornamental and timber trees native to the swamps of southern North America, though readily adaptable to cooler climates. Although *Taxodium distichum* trees are generally deciduous, leaves persist year-round if grown in a warm climate. A tree growing in periodically flooded habitat produces conical woody projections, or knees, from the lateral roots. The needle-like flat leaves are alternately arranged in two ranks along small twigs. The nearly spherical seed cones mature and break apart in one season. *Taxodium distichum* is native to the American Southeast, and one variety, *Taxodium distichum* var. *imbricatum*, is native to its coastal regions. Commonly called the pond cypress, it is generally smaller and narrower than the species, usually growing on the margins of ponds and lakes rather than in water. Some taxonomists consider it a distinct species, *Taxodium ascendens*. The closely related *Taxodium mucronatum*, or Montezuma-cypress, is native primarily to Mexico and Guatemala, with foliage like the bald-cypress (*Taxodium distichum*) but with somewhat smaller cones and shorter leaves, retained year-round in its native habitat.

Juniper (*Juniperus*) has by far the greatest number of species and the broadest geographical and ecological range of any genus in the cypress family. Primarily distributed in the northern hemisphere, including North America, Europe, and Asia, it is also found in several semiarid regions, including northern Mexico, western United States, and central and southwest Asia. *Juniperus* is the only genus found in both hemispheres; *Juniperus procera* ranges from Saudi Arabia through Ethiopia to Zimbabwe. Junipers have a wide variety of growth habits, from low prostrate shrubs like *Juniperus horizontalis* to large trees like *Juniperus virginiana*. They are characterized by their berry-like seed cones with seed scales of varying textures, from fleshy to woody. Initially small and green, they mature to dark blue or dark purple and occasionally red-brown. This adaptation of small, berry-like cones allows seed dispersal by birds rather than the wind. Most are dioecious, though some species are monoecious.

Dawn redwood (*Metasequoia glyptostroboides*), the sole species in the genus *Metasequoia,* is a deciduous conifer first discovered in 1941 in Sichuan, China; fossil records indicate that they have existed for 50 million years. Dawn redwoods develop into large, tall, conical trees with crowns that broaden with age. The reddish-brown, exfoliating bark becomes deeply fissured and fluted with age, forming a very wide trunk at the base, with pronounced root flares called buttresses. The green foliage is soft, linear, and

A beautiful dwarf variety of dawn redwood *Metasequoia glyptostroboides* 'Miss Grace'.

fern-like and turns an attractive reddish-brown before dropping in the fall. The stems and needles grow in opposite pairs. The needles fall together with the attached branchlet. Numerous pollen cones about ¼-inch (0.64-cm) long are arranged in long spikes or panicles. The small (¾-inch, 1.9 cm) brown seed cones hang pendulous on sparsely leaved lateral branchlets.

Giant Sequoia (*Sequoiadendron giganteum*), the only species of the genus *Sequoiadendron,* is native to the Sierra Nevada range of California and the largest of all trees by volume. It is a massive pyramidal evergreen tree with reddish-brown, fire-resistant bark. The scale-shaped or awl-shaped leaves grow tightly against the branches. The small (⅕-inch, 0.45 cm) pollen cones are on the same terminal branch as the seed cones. The 2-inch (5.48-cm) long seed cones mature in two seasons and open following a wildfire.

Coast redwood (*Sequoia sempervirens*) is restricted to a narrow strip about 20 miles wide along the typically humid western coast of the United States, from southern Oregon to Monterey, California. The redwood is the world's tallest tree and can live for more than 2,000 years. It has a straight, erect trunk, often buttressed at the base, with thick, fire-resistant, deeply fissured, fibrous, peeling, reddish-brown bark. It has the unusual ability to regrow by sprouting from a stump of a tree destroyed by fire. The leaves are variable—long and flat on young trees and shaded shoots of old trees but scale-like on shoots in full sun on older trees. The leaves are spirally arranged on the shoots, but the larger, shaded leaves are twisted at the base in a flat plane to capture the maximum amount of light. Leaves have two broad whitish stomatal bands underneath. The small solitary pollen cones, about ⅙ inch (0.42 cm) long, are at or near the ends of buds where leaf and branch meet. Seed cones about 1 inch (2.54 cm) long grow at the end of short branchlets.

Incense-cedar (*Calocedrus*) is native to western North America and eastern Asia. The species *Calocedrus decurrens* is distributed throughout western North America. The tree has a single straight trunk with smooth young bark that becomes fibrous, furrowed, and peeling with age. The densely branched tree has short horizontal branches starting at the base. The tree form is narrowly columnar but eventually broadens with age. The flattened branchlets are covered with fern-like sprays of scale-like foliage in apparent whorls of four, but the leaves cross each other, with the adult lateral leaves folded over their bases, overlapping the flat facial leaves. Pollen and seed cones form at the ends of different pendulous branchlets. The monoecious cones mature in a single season.

Arborvitae (*Thuja*) is an evergreen tree with stringy-textured reddish-brown bark peeling in longitudinal narrow strips. The small branches grow in fan-shaped, flattened sprays with aromatic scale-like leaves in opposite pairs in four ranks. Small pollen and seed cones are borne at the ends of short branchlets. The seed cones mature in a single season. Among the six species of *Thuja,* two are widely planted in horticulture: the western red-cedar (*Thuja plicata*) distributed in western North America, and the eastern arborvitae (*Thuja occidentalis*) native to eastern North America.

This iconic coast redwood (*Sequoia sempervirens*) graces the author's Puget Sound garden.

Brilliant red arils on a golden yew.

Family: Yew (Taxaceae)

These primarily dioecious small to medium evergreen trees and shrubs are distributed throughout much of North America, Asia, Central America, Europe, and northwest Africa. The yew family's dark, flattened, linear to lanceolate needle-like leaves with stomata bands on the undersides are attached singly or in pairs in a spiral arrangement but are most often twisted at the base to appear two-ranked and spread out horizontally on either side of the twig. The pollen cones shed pollen in early spring. Unlike other conifer families, members of the family Taxaceae have seed cones that bear a single seed surrounded by an aril—a fleshy, usually brightly colored, soft, sweet, and juicy covering. Birds eat the seeds and disperse them, undamaged, in their droppings; squirrels bury them to keep for the winter—the uneaten ones then germinate. Two genera are commonly cultivated in our landscapes: the true yews (*Taxus*) and the nutmeg-yews (*Torreya*, or Torrey pine).

True yews (*Taxus*) grow primarily in the temperate latitudes of the northern hemisphere in Europe, North America, and Asia, with some species found in the tropical highlands. They are relatively slow-growing evergreen trees and shrubs with flat, dark-green leaves in two flat rows on each side of the stem. They are widely used in ornamental horticulture; however, the leaves, pollen, and seeds of all true yew species are highly toxic to humans due to the presence of the alkaloid taxine, and they have a high allergic potential. Most yews are dioecious. The seed cone contains a single seed partly surrounded by an aril, which matures in about 18 months.

Nutmeg-yews (*Torreya*) are native to Asia and North America. A genus of six species, one species is native to California, one to Florida, three to China, and one to Japan. The Japanese species, *Torreya nucifera*, has an edible seed. Nutmeg-yews are small to medium evergreen trees with linear, bristle-pointed leaves, spirally arranged on the shoot and twisted at the base such that they lie in two flat ranks. The small pollen cones hang along the undersides of the shoots. The seed cones mature in about 18 months, forming a simple nut-like seed fully surrounded by an aril.

Family: Plum-yew (Cephalotaxaceae)

Plum-yews (*Cephalotaxus*) are a single genus of eight species native to eastern Asia. Once considered part of the yew, or Taxaceae, family they are now recognized by most taxonomists as a separate family. They are evergreen shrubs or trees, either dioecious or monoecious, with the pollen cones and seed cones generally formed on different branches. The seed cones are borne singly or grouped on short stems, maturing over a long period into a single large nut-like seed fully enclosed by an aril. Unlike members of the yew family, the plum-yews have edible seeds. As in the yew family, the leaves are spirally arranged on the shoots but twisted at the base and in two flat ranks on the stem; the linear, soft leaves have a blunt tip. Several species are cultivated widely, including the Japanese plum-yew (*Cephalotaxus harringtonii*) and the longleaf plum-yew (*Cephalotaxus fortunei*).

Family: Japanese Umbrella-Pine (Sciadopityaceae)

The Japanese umbrella-pine (*Sciadopitys verticillata*) is the sole species in this family. It gets its name from the whorls of "leaves" at the tips of the branches; some botanists consider these modified stems, while others consider them to be pairs of leaves fused together. This leaf-like stem tissue, up to 5 inches (12.7 cm) long, is flexible, flat, dark green, and glossy. The "leaves" persist for three to four years and perform photosynthesis just as true leaves do. Japanese umbrella-pines are very slow-growing, dense, narrow conical to pyramidal evergreen trees with heavy branching. The trees, with their luxuriant foliage, typically mature in cultivation, after many years, to 20 to 30 feet (6.1 to 9.1 meters). Branches occur predominantly in whorls of three to five. Pollen cones occur in dense clusters at the top of branchlets. The single upright seed cones with flat scales mature in about 18 months and remain intact but open to release the seeds.

Japanese umbrella-pine (*Sciadopitys verticillata*) shown in early spring backed by a blooming cherry.

Family: Podocarp (Podocarpaceae)

This large family of monecious or dioecious evergreen shrubs or trees is primarily naturally confined to southern hemisphere island regions, especially New Zealand, New Caledonia, and Tasmania. The family also occurs in South America and Africa, with a significant number north of the equator in Southeast Asia, just reaching the most northerly point of. These conifers tend to have horizontal branches and flattened, often broad, leaves attached individually in a spiral arrangement or in pairs arranged in flat rows on either side of the horizontal branches. This adaptation permits them to harvest light efficiently in shaded environments and may explain why they can prosper in wet tropical forests. It also allows them to thrive in shady areas of the garden. The pendent seed cones have just a few fleshy bracts or scales and mature in a single season. The seeds, which are dispersed by animals, are covered by a fleshy outer scale structure called an *epimatium*. The most widely distributed and perhaps the best known genus in the family is Podocarp (*Podocarpus*), which includes about a hundred species in Africa, Asia, Australia, Central and South America, and several Pacific Islands. They are evergreen trees and shrubs with simple, often lanceolate, flattened leaves with a distinctive midrib, usually spirally arranged on the stems, though the leaves of some species are twisted and appear in two horizontal ranks. The juvenile leaves are usually larger than those on mature plants but similar in shape. Most podocarps are dioecious, with the long, catkin-like pollen cones and the mature, brightly colored, fleshy seed cones borne on separate plants.

Family: Celery-Pine (Phyllocladaceae)

The phylloclades or celery-pines (*Phyllocladus*) are four species of a single genus native to Malesia (a floristic region comprising Malaysia, Indonesia, the Philippines, and Brunei), New Zealand, and Tasmania. Evergreen trees or shrubs with single upright trunks or multiple trunks with smooth or scaly bark, they are distinguished by having *phylloclades* (meaning "leaf stem")—flattened branches that look like and function as photosynthetic leaves. The resin is in the bark and phylloclades. The young phylloclades are spirally arranged or in pseudo-whorls on long shoots, becoming scaly as they mature. The celery-pines are usually dioecious but occasionally monoecious. Pollen cones are spirally arranged in clusters near the end of short shoots, while seed cones are borne singly or in clusters of two to four. The seeds are covered by a fleshy or leathery cup-like aril. The celery-pines have a distinctive and attractive appearance, yet they are seldom cultivated outside of Australia and New Zealand. They deserve a place in sustainable gardens elsewhere in the mild regions of the world.

Family: Araucaria (Araucariaceae)

The monkey puzzle tree (*Araucaria araucana*) and Norfolk Island-pine (*Araucaria heterophylla*) are probably the species best known by American gardeners. These dioecious or monoecious tall evergreen trees are native to forested regions primarily in the southern hemisphere, including New Caledonia, New Zealand, Australia, and South America. The sharply pointed, leathery leaves can be needle-like or broad, ranging from linear or oblong to awl-shaped, scale-like, claw-shaped, or broad-bladed as in the genus *Agathis*, arranged spirally or in crisscross or overlapping pairs. Regular horizontal branching with a tiered appearance is typical for this family. Pollen cones with many spirally arranged pollen scales form in leaf axils at the tip of branchlets or the end of side shoots. The large, round seed cones contain one seed per scale and are usually borne upright at the tips of branches, maturing in two years and then shattering.

The newest member of this plant family, *Wollemia nobilis*, was discovered in 1994 in the Blue Mountains of Wollemi National Park, northwest of Sydney, Australia, and is now available in cultivation. Previously, it was known only from fossil records dating back 200 million years. Unlike many of the other members of this plant family, it has quite soft leaves. The juvenile leaves are dark green and fern-like with waxy undersides; the adult leaves are yellowish-green, stiff, flat, and linear. The branches are unusual as the side branches seldom have further branching, and, after a few years, they stop growing and some terminate in a cone. Once the cone on the side branch matures, the side branch dies.

With all this conifer information in mind, it is time to plan the new or transformed garden landscape you envision, with conifers taking their rightful place. This requires making a thorough evaluation of your existing garden—the focus of the next chapter.

Imagine your ideal garden.

Evaluating Your Garden

Chapter 3

When many of us imagine a garden, we picture one that is vibrant and attractive, with a riot of bright colors, delightful fragrances, and happy hummingbirds. With that image in mind, who wouldn't want to simply walk into a garden store and grab the flashiest trees and the most outrageous flowers? But the gardener gossip grapevine is full of tales of inspired novices forgoing planning and having to learn some hard truths when their efforts fail to manifest the garden of their dreams. To avoid this fate, begin with a foundation of conifers and the subtler, though no less awe-inspiring qualities they bring.

A sustainable garden endures over time without excessive expenditure of resources or maintenance hours while enhancing biodiversity and creating healthy habitats for wildlife. Ideally, an enduring garden mimics nature and allows natural processes to continue to shape the landscape. This means that the right conditions must be created to allow such processes as the nutrient cycle (including photosynthesis and food and decomposition webs) and the water cycle (which plays a part in sediment transport and soil formation) to continue. Evergreen, deciduous, herbaceous, and woody plants are a necessary part of

You may find it helpful to create a detailed site map before you start making changes to your garden.

the diversity and richness of the landscape. As such, they should be considered a necessary part of your garden plan, which weaves human needs and uses with the needs and uses of nature—because, at the end of the day, these are one and the same. Plants selected for the garden should meet the goals of ecological diversity, natural disease and pest resistance, and environmental sustainability while minimizing negative impacts for the environment or people. Confers fit very well into a sustainable landscape because they are long-lived, drought-tolerant, adaptable to adverse conditions, and provide a wealth of ecological services to both us and wildlife.

Wander the aisles of your favorite local plant nursery all you like, getting to know each conifer face-to-leaf, but be patient and hold off on making the big investment of time, money, and elbow grease until you have studied your landscape in its current form and done your research on potential conifers and companion plants and their care. Think of this time and energy invested in preparation as the gardener's version of "measure twice, cut once." Also, the better you know your landscape, the easier it will be for you to choose appropriate conifers and develop a more sustainable garden.

Wander the aisles of your local nursery and compare. Take your time and make notes before buying.

Climate

You will first need to know the climate and microclimates at your site. Climate is the average weather conditions prevailing in a region over an extended period. This includes seasonal variations in temperature, humidity, precipitation, sunshine and cloudiness, day length, wind, and barometric pressure. Because climate determines precipitation patterns, soil temperature, and moisture content, you need to gather this information in order to choose appropriate conifers. At the very least, you need to know the minimum low temperatures in winter, the highest temperatures in summer, and the expected amount and seasonal variation of rain or snow.

A conifer's ability to grow and survive in a region largely depends on its *hardiness*—its ability to withstand a minimum low temperature. Therefore, the USDA has developed a cold hardiness zone guide (https://planthardiness.ars.usda.gov/PHZMWeb/) based on average minimum winter temperatures. The lower the zone number, the lower the minimum temperature range. For example, Fairbanks, Alaska, is in zone 1, with minimum annual temperature of below –50°F (–45.6°C); Honolulu, Hawaii, is in zone 11, with a minimum annual temperature above 40°F (4.4°C). If you live in zone 5a, the average annual extreme minimum temperature is –20° to –15°F (–28.9° to–26.1°C). If a conifer is known to be hardy to zone 6a (–10° to –5°F; –23.3° to –20.5°), you would be taking a big chance of losing one that you plant in zone 5a. Whether a particular conifer can be grown in a zone depends on whether it can withstand the zone's minimum temperature.

The number of days per year that a region has high temperatures is also an important consideration for the growth and survival of most of our cultivated conifers. For gardeners to evaluate the potential impact of high temperatures on plant growth, the American Horticultural Society has developed the Plant Heat Zone map (http://solanomg.ucanr.edu/files/245158.pdf) based on the average number of days per year that temperatures climb above 86°F (30°C). There are 12 heat zones, ranging from zone 1, with less than one day per year above 86°F (Alaska, central Colorado), to zone 12, which has temperatures above 86°F for more than 210 days per year (the southern tips of Texas and Florida). While the heat zone map may be helpful, the absolute temperature above a certain level may also be critical. So you should also determine what the highest temperature is likely to be for your area.

For a list of species that can tolerate the combination of heat and humidity and those that can tolerate warm and dry conditions, see the list in Chapter 5, "North American Regions Where Conifer Species and Their Cultivars Can Be Grown."

Both of the maps just mentioned are helpful, but it is important to recognize that they should be used only as general guides; they do not take into consideration the timing of cold or heat spells, the year-to-year variation in minimum and maximum temperatures, rain and snow patterns, wind, soil moisture, humidity, and amount of sunlight—all of which have an impact on a plant's ability to withstand extremes in temperature.

Furthermore, whether a conifer can withstand a low temperature very much depends on whether the plant has undergone winter acclimation or is dormant, as actively growing conifers are much more susceptible to low temperatures. Furthermore, as the climate changes, conditions are becoming more unpredictable; there is greater variability in winter and summer temperatures, and hardiness zones are shifting. Take note, though, that although the climate is warming, it is best not to expose a conifer to a hardiness zone lower than it is rated.

Where in your garden you plant your conifer will depend on the *microclimate*, which refers to a specific garden site that differs from the surrounding general area. An important consideration is the orientation of the site toward the sun, which affects the angle of the sun's rays and the hours of sun per day hitting the area throughout the seasons. For example, a south-facing hillside would receive considerably more sun and be warmer than a north-facing hillside. The proximity of the site to buildings and other structures or vegetation can provide shading for part of the day, which would affect how much sun and wind the site receives. If the garden has a higher elevation than the surrounding area, it may be colder and windier than the surrounding area. If the garden is on a slope, that will affect the amount of wind the area receives and increase the movement and rate that water drains on the site and the likelihood of erosion. Proximity to water can moderate temperatures and affect wind speed. Good drainage is a key factor in determining the successful growth of most conifers. All these elements contribute to a microclimate.

Soil

The soil serves as the medium for conifer growth. As such, it provides vital nutrients, minerals, and gas exchange to the conifer's roots. It provides structural support, as well as habitat for trillions of beneficial soil organisms. The soil protects the conifer from the effects of erosion, while holding water and making it available for conifer growth. During the process of water infiltration, the soil filters the water and removes harmful chemicals and other pollutants. The soil helps regulate the amount of atmospheric carbon dioxide by storing carbon in the form of stable organic matter complexes. The type of soil can make a big difference in conifer health and growth; conifers do best when the soil has the appropriate nutrients and pH, along with a loose and granular structure.

The relative proportion of small clay particles, intermediate silt particles, and large sand particles determines your soil texture, which, in turn, largely determines the soil's water infiltration rate, water-holding capacity, nutrient retention, and drainage—all properties that are important for healthy conifer growth. Although clay soils usually have excellent nutrient retention and water-holding capacity, they are prone to compaction and tend to limit water infiltration. In contrast, sandy soils have excellent porosity and aeration but have poor nutrient retention and water-holding capacity. Silt has intermediate properties between those of clay and sand particles. To get a general sense of your

This photo shows healthy, loamy soil with the appropriate proportion of sand, silt, and clay particles to allow for both good water-holding capacity and good drainage.

soil's texture, add water to a handful of soil until it is moist throughout. If the soil feels gritty and crumbles, it has a high content of sand. A sticky and slippery soil indicates a high content of clay. If the soil feels smooth and greasy, then there is a lot of silt in the soil. A healthy soil loam usually contains about an equal proportion of silt and sand and somewhat less clay—anywhere from 8 to 30 percent. Consequently, loam has near-optimal properties for conifers in terms of drainage and water and nutrient retention. If your soil has too much of one or the other component, do not despair—and do not go out and buy multiple bags of what your soil is missing. It is not practical to try to significantly alter your soil's texture. Instead, pick a conifer that can thrive in the soil you are able to give it. Many conifers can adapt to a less-than-ideal balance of soil particles.

Soil pH

Soils vary in their relative acidity or alkalinity, which is referred to as *pH*. The pH scale goes from 1 to 14, with the numbers below 7 considered acidic and the numbers above 7 alkaline, or basic; a pH of 7 is considered neutral. Because pH is an important determinant of how well conifers can use nutrients from the soil, you should know your soil's pH before making your conifer selection. Most conifers and other plants are well adapted to acidic soil in a pH range from 5.5 to 6.5.

Soil Testing

To accurately determine your soil's pH, organic matter content, concentration of various nutrients, and other properties, get your soil tested by a certified soil testing facility. Local County Extension agents or conservation departments can provide information on soil testing facilities. The results will guide not only your plant choice but your nutrient management in the future.

Soils can vary even within a single garden. Therefore, you should obtain soil samples from different areas of your garden, particularly the areas in which you are considering placing a conifer. The soil testing facility you choose can provide you with explicit instructions on how to take a soil sample. For more information on what to do about soil pH, see Chapter 8, "Sustainably Caring for Conifers."

Soil Compaction

Compacted soil can restrict root growth, reduce the capacity of the soil to hold water, disrupt fungal associations with conifer roots, and reduce nutrient, water, and oxygen uptake by roots. Hilltops exposed to wind and rain are more likely to have shallow soils and tend to be nutrient deficient, whereas low spots will generally have deeper and more nutrient-rich soils. Assuming you don't have very rocky soil, one straightforward way to estimate soil compaction is to determine how far you can push a soil probe or piece of rebar into the soil. If the bar goes into the soil only a few inches, the soil is compacted; if it goes in 4 to 6 inches, it is moderately compacted; if it goes into the soil 8 inches or more, there is little compaction. Chapter 8 also provides information on what to do about soil compaction.

Soil Drainage

Most conifers prefer cool conditions and relatively fast-draining soils. To assess your stormwater flow and drainage patterns, wait for a rain and then look for runoff routes and pools. An abundance of impervious surfaces can contribute to excess surface water flow; the tendency of water to pool in low-lying areas signals poor drainage and possible soil compaction. You might also find evidence of erosion, such as narrow and shallow channels on the soil surface or soil deposits at the bottom of slopes; these suggest compacted soil and/or excessive stormwater runoff. You can get a quantitative assessment of drainage by doing a percolation test:

You can assess your drainage by doing a simple percolation test.

1. Dig a hole 1 foot in diameter and 1 foot deep.
2. Fill the hole with water several times, letting it drain between each filling.
3. Refill the hole to the top and allow the hole to drain a few inches. Measure the height of the water.
4. Recheck the height of the water after 2 to 4 hours.

If the soil drains between 1 and 3 inches per hour, the drainage is reasonable. If it drains less than 1 inch per hour, drainage is poor, and efforts to improve drainage should be made. If the soil drains 6 inches or more per hour, the drainage is too fast. For more information on what to do about poorly draining sites, see Chapter 8.

Sunlight

How the different areas of your garden are orientated to the sun and whether buildings and vegetation create shading are factors directly tied to microclimate temperature. During winter in the northern latitudes, the sun rises in the southeast, sits low in the southern sky at noon, and sets in the southwest. In summertime, the sun rises in the northeast, sits high in the southern sky at noon, and sets in the northwest. Most conifers prefer cool conditions and full sun, but many do well in partial shade or filtered sun, and some perform well in light shade (see "Areas with Light Shade or Dappled Shade" in Chapter 5). High-intensity sunlight can damage leaves and cause sunscald in some conifers, particularly in winter. In order to select the conifer that is right for your garden, you will need to determine the number of hours of sunlight each zone receives throughout the day and through the seasons.

Frost

You should be aware of which areas in your garden are prone to frost. Low areas are more vulnerable because chilly air flows downhill and settles in low spots.

Slope

Slope faces tend to be warmer than both hilltops and valleys because heavy, cold air mixes with warmer air on its downward trajectory. Low areas are more prone to frost and colder temperatures because chilly air flows downhill and settles in low spots. If the conifer you selected is of borderline hardiness, it's best not to place it in a low spot in your garden. The crests of slopes, meanwhile, receive more wind than valleys, and proximity to a body of water increases wind exposure. Conifers are particularly susceptible to the drying effects of a cold wind, and broken branches and trunks are more likely to occur on a windy slope. You can reduce damaging impact by planting a windbreak. For more information on windbreaks, see Chapter 4.

In the next chapter, we'll explore the range of specific roles for conifers in the garden, including many specialty cultivars and settings to suit every need and taste.

Designing with Conifers

Chapter 4

Once you've evaluated your site conditions, the next step is to set some clear goals for the additions and changes you wish to make to your landscape. The conifers you select should not only provide year-round interest and color but should also serve specific functions. How would you like your garden to be used? How can your conifers make your garden more functional, comfortable, and attractive? By exploiting differences in heights and shape and thoughtfully combining complementary colors and contrasting textures, you can create a garden that looks good, makes you feel good, and provides important ecological services.

The possibilities can seem overwhelming, so it helps to have some parameters. You can start by deciding what role the conifer will play in your garden. Consider the following uses for conifers:

- Hedges and screens
- Windbreaks
- Accents and focal points
- Structure or backbone
- Multilayered property border
- Groundcover
- Support for spring and summer climbing vines
- Container plantings
- Framing for a special vista
- Shade providers for hardscapes
- Complements to structures and other plants in a Japanese-style garden

Slow-growing or miniature conifers are ideal for rock gardens.

Remember that the key step in any planting plan is to match each plant to the specific conditions of the planting site. I suggest you start by making out a list of what you would like to include in your garden design and then proceed to draw a site plan. Use simple shapes, such as circles or ovals, to indicate the placement of the general elements you want in each garden area. After you have developed your ideas in the form of this "bubble diagram," you can then develop a more detailed plan that will allow you to decide which specific plants you want in each "bubble." Then, you can generate a specific planting plan.

Once you have determined the use for your conifer, the next steps are deciding on tree size, form, color, and texture.

Size

You will want to select a size of conifer that fits the scale of your garden, but be aware that the growth rate and ultimate size (width and height) of your conifer can be considerably different from the rate and size predicted by the nursery. (Note that the estimates nurseries give on a plant's label are usually for the tree's size at ten years). The specific soil characteristics, the sun exposure, temperatures throughout the day and seasons, and the amount of moisture the plant receives (particularly during its annual flush of growth) all have an important influence on growth rate. (For information on conifer care for optimal growth, see Chapter 8.)

Conifers are designated in the trade by these *growth rates*:

- Miniature (M): <1 inch/year
- Dwarf (D): 1 to 6 inches/year
- Intermediate (I): 6 to 12 inches/year
- Large (L): >12 inches/year

The Oriental spruce *Picea orientalis* 'Aureospicata' with its unique new growth and colorful cones.

Keep an Eye on Growth Rate

Many conifers that are transplanted from a nursery take a few years to get established before they start to grow at a consistent rate. Once your conifer appears to be healthy and established in its new home, measure the growth rate every few years to calculate the average. If you observe that the growth rate is greater than anticipated, and it appears that ultimately the conifer will not have adequate space, it's best to transplant it to a more appropriate location before it gets too big.

Slow-growing dwarf or miniature conifers are ideal for small gardens, such as rock gardens. *Intermediate and large conifers* work well as privacy screens, windbreaks, or shelter belts on your property line, or as structural background. *Narrow and tall conifers* can serve as accents in either small or large gardens, as they contrast well with a garden's low-growing horizontal components. You will want to provide generous spacing to allow for good air circulation and future growth. Keep large conifers away from walkways and the foundation of your home, and make sure they won't eventually block desirable views.

Form

To choose the form—that is, the shape and size of a mature conifer and how the branches are arranged on the trunk—you must consider the garden role you're giving the conifer and how it will blend with your other garden plants. Usually, it's best to select a conifer with an even and symmetrical distribution of proportioned branches, although one with an irregular habit may be suitable as an accent.

Conifers come in a wide variety of shapes. Here are just a few of them:

- Ground-hugging pancakes, such as *Juniperus rigida* var. *conferta* 'Emerald Sea,' *Juniperus procumbens* 'Nana,' and *Microbiota decussata.*

The groundcover Russian-cypress (*Microbiota decussata*) has a distinctive brownish-bronze color in winter.

- Globes or round mounds, such as *Thujopsis dolabrata* 'Nana', *Chamaecyparis lawsoniana* 'Rimpelaar', and *Abies lasiocarpa* 'Cvikov'.

The globe-shaped Hiba arborvitae (*Thujopsis dolabrata*) does well in both sunny and shady conditions.

- Cone-shaped, such as *Abies koreana*, *Abies nordmanniana*, and *Abies alba* 'Pyramidalis'.
- Columnar, such as *Taxus baccata* 'Standishii', *Thuja occidentalis* 'Joe's Pillar', and *Cupressus sempervirens*.

Be aware that the overall shape of a conifer often changes as it matures. A young conifer may start out as a round mound and eventually develop a leader and become a small tree, or it might begin as an upright pyramid and ultimately spread into an open, flat-topped specimen. Or the tree may have an upright growth pattern with the lateral branches weeping.

The European silver fir (*Abies alba*) 'Pyramidalis' forms a perfectly symmetrical cone.

The Italian cypress (*Cupressus sempervirens*) forms an exclamation point in the landscape.

Color and Texture

Left: The Japanese-cedar 'Elegans' shows off its leaves with a purple hue in winter.

Center: The mugo pine 'Amber Gold' shines a bright yellow in winter.

Conifers offer a wide spectrum of colors and a varied assortment of textures. For most of the year, the color of the foliage—in myriad shades of green, blue, gray, silver, yellow, maroon, purple, orange, and brown—is the dominant attraction, while the texture of the foliage and the bark or cones adds a layer of beauty and nuance. Gold brightens up the winter landscape and shady corners of the garden, while the grayish color of the waxes and resins in the leaves helps harmonize the many different garden colors. For a list of conifer cultivars that are golden/yellow in winter and green in summer and those that are golden/yellow year-round, see Table 1 and Table 2 in the Appendix. The tables also include other characteristics of these trees. Note that care should be taken with the

placement of gold or variegated plants due to their susceptibility to foliage burn in full sun and warm climates.

Many conifers change colors throughout the seasons, usually in response to changes in temperature. Often, the new conifer foliage in spring is lighter in color than the old foliage. For example, *Cryptomeria japonica* 'Elegans' goes from a soft green in summer to show-stopping purple in winter. The cultivars *Pinus mugo* 'Amber Gold' and *Pinus contorta* var. *latifolia* 'Chief Joseph' turn from an attractive green in spring and summer to a distinctive gold as cold weather arrives, returning to green again when the weather warms up in spring and new growth arrives. And the new growth of *Pinus contorta* var. *latifolia* 'Taylor's Sunburst' emerges golden in the spring.

The spring growth of the lodgepole pine 'Taylor's Sunburst' contrasts with the older green needles.

The autumn leaves of the golden-larch (*Pseudolarix amabilis*).

The leaves of deciduous conifers undergo a striking color change in the fall just before leaf drop, from green to a golden bronze or coppery color. Among the many deciduous conifers in my garden, I find that the golden-larch (*Pseudolarix amabilis*) has the most striking, soft, feathery, golden foliage.

There are many cultivars with variegated foliage; some are bicolored (having two colors on separate stems of the same plant), and others have different colors on the same leaf or needle in the form of stripes—such as *Pinus densiflora* 'Golden Ghost' and *Pinus cembroides* 'Pina Nevada'—or blotches or spots, such as *Chamaecyparis lawsoniana* 'Ellwood's White'. Variegated foliage can provide a lovely contrast in an otherwise all-green landscape.

The variegated needles of Japanese red pine *Pinus densiflora* 'Golden Ghost'.

Conifer *cones* change colors too as they develop, going through a variety of decorative color stages, some of which are very showy. For example, the Korean fir has bright violet or blue new cones, and those on *Picea abies* 'Pusch' are carmine; and the seed cones of junipers and yews often have brightly colored, fleshy, berry-like coverings that are an attractive contrast to their foliage.

The green foliage of yews contrasts with the bright red, fleshy seed covering, called an aril.

The summit athrotaxis *Athrotaxis × laxifolia,* is an excellent example of a conifer with coarse texture.

Both foliage and bark, along with how the branches are arranged, can influence texture. The summit athrotaxis *Athrotaxis × laxifolia* has coarse foliage. In contrast, *Taxodium distichum* 'Cascade Falls' (weeping bald-cypress), *Tsuga canadensis* 'Pendula' (weeping Canadian hemlock), and *Cedrus deodara* 'Cream Puff' have fine foliage.

Also consider the many colors and distinctive textures of *bark*. Some examples of conifers with striking bark include

- *Abies squamata*, a fir with flaky bark.
- *Pinus parviflora* 'Ara kawa,' with rough bark (prized by bonsai enthusiasts).
- *Pinus thunbergii* var. *corticosa* (Japanese cork bark black pine), used extensively in bonsai.
- *Metasequoia glyptostroboides* (dawn redwood), with its particularly attractive russet-colored peeling bark and fissured and buttressed trunk.

The deodar cedar 'Cream Puff' is an excellent example of a conifer with fine texture.

General Landscape Considerations

After you've evaluated your site and understand the local conditions, here are the next steps:

- Choose plants that will thrive in your microclimates and, hopefully, those that can adapt to a change in climate or conditions.
- Take into consideration each plant's mature size, longevity, and competitiveness, and its scale in relation to the chosen site and other plants in the community; you want the relative proportions of the plant groups to be in proper balance. The plants in your mix should compete minimally for space and resources.
- Consider each plant's maintenance and resource requirements over time.
- Before planting, prepare the soil in the entire planting area; make sure the site is free of weeds, and ensure the soil is well-draining and rich in organic matter.
- Lay out and plant your conifers first, as they will be your key structural elements.
- Place your plants in a staggered arrangement in terms of form, color, and height, rather than in a single row, although the tallest should be in the back, and the smaller ones should primarily be toward the front of the planting area.
- Plant the smaller plants—perennials, grasses, and bulbs—in natural drifts or in sweeps, with wider spacing at the edge of the grouping, so they appear as if they were self-seeded.

This group of conifers exhibits an attractive staggered arrangement.

The Serbian spruce 'Pendula Bruns' rises up out of a hillock of heather *Calluna vulgaris* 'Spring Cream'.

- Add a top layer of mulch to reduce weed germination and enrich your soil, but keep the mulch a few inches away from the trunks of trees.
- Carefully monitor your plants while they get established; make necessary adjustments; and promptly deal with any problems that arise. (For more details about soil preparation, planting, and plant care, see chapters 3, 6, and 8.)

Because conifers grow slowly, it's best to fill in the space around each conifer with groundcovers, perennials, or grasses that can be easily removed and transplanted as the conifer grows. When transplanting these "filler" plants, be careful to avoid disturbing or damaging the conifer's roots; take care not to allow these filler plants to crowd or shade the lower portion of your conifer (especially if it is shade intolerant), as this can cause unsightly dieback of the conifer's lower foliage. These fillers should add visual interest, blend well, and accentuate the shape and color of your conifer so that your conifer will shine in the landscape, regardless of season.

Uses of Conifers in Urban Gardens

The versatile conifer can be used in almost any garden design. Whether you are looking for a groundcover to replace some of your lawn, a screen to hide your neighbor's garbage cans, a background for a bed of flowers, or a stunning addition for an island bed or mixed border, a conifer can be found to suit your purpose and shine through all seasons. Let's examine the many potential uses of conifers.

Hedges and Screens

Fences and walls made of brick, wood, and chain link are common in residential neighborhoods. They provide privacy, keep out stray animals (and people), and hide unsightly views, but they can also appear uninviting and obtrusive, even unfriendly. In my opinion, a living screen or hedge is often a better alternative because it can serve the same function as a fence—such as creating a sense of seclusion or defining a property line or garden border—while serving multiple other functions: filtering air and water pollutants; providing food, protective cover, and nesting sites for birds and other wildlife; and enhancing the diversity of plant forms, colors, textures, and species that add beauty to your landscape.

When considering a hedge or screen, the first step is to decide what specific function it will serve. (If designed well, screens can serve multiple purposes.) How high do you want it to be? Does the foliage need to be dense? High or low maintenance? For shady or sunny conditions? What are your preferences for color, texture, and overall aesthetics? Do you want the screen to be formal or informal?

The more traditional formal screen is composed of a straight row of one species or cultivar, usually uniform, with a clearly delineated geometric shape. These are the types of hedges you might find in the gardens of European royalty or outside the mansions

Sheared yew hedge along a path in Kubota Garden in Seattle.

of New England. Though these are doable by amateur gardeners, nowadays most people prefer informal screens that have a more naturalistic aesthetic, with more freeform placement and plant variation. An ideal informal screen consists of a variety of species or cultivars, including mixed plantings of conifers or a combination of conifers and long-lived broadleaf evergreens. Both mixes can provide more ecological benefits in a natural setting than a single row of the same plant. From a design perspective, using the same slow-growing, long-lived conifer (or conifers with similar characteristics) throughout a garden can help unify a garden and provide a sense of continuity and permanence; this can be of particular value in a garden that is a mix of perennials and other herbaceous plants. Screens are nearly always best when designed with mixed evergreen plantings, as they provide a vibrant and sustainable privacy screen year-round. Either way, for the purpose of screening, it's important to select cultivars with dense canopies that extend all the way to the ground, while also factoring in climatic conditions.

For a formal screen or hedge, yews tend to be the best option. They make an excellent dense, long-lived hedge, as they grow back readily when pruned (if not cut back too far), even on mature wood, and they take full sun or light shade. Start with young plants planted close together (about 2 feet apart); you will shape and control the width and height of your hedge with regular trimming as it grows. There are several different yews to choose from for a hedge:

- For warmer areas, with a minimum cold hardiness of USDA zone 6 (a low temperature of –10°F to 0°F), plant male or female English yews (*Taxus baccata*) or Irish yew (*Taxus baccata* 'Stricta'), a relatively narrow female clone that bears fruit.
- For colder areas, with a minimum cold hardiness of zone 4 (–30°F to –20°F), plant Japanese yew (*Taxus cuspidata*) or Hicks' Hybrid yew (*Taxus* × *media* 'Hicksii'), a cultivar of the hybrid between the English yew and Japanese yew.

Narrow, upright junipers such as *Juniperus virginiana* 'Taylor,' *Juniperus scopulorum* 'Woodward,' and eastern arborvitae (*Thuja occidentalis*) are other good choices for cold-weather conifer hedges.

Proper soil preparation and planting will ensure the successful establishment of your hedge. If your soil is not rich in organic matter, incorporate high-quality fine compost into the entire hedge planting area. Yews prefer organic-rich, well-draining sandy loam. As the hedge grows, trim the sides but do not trim the top until it reaches or exceeds the desired height. Taper the sides to make the hedge wider at the bottom, so that more light can reach the bottom branches.

An unsheared border of yews can make an eye-catching low-barrier planting.

For an informal screen/hedge, base your choice on interesting colors, textures, and shapes. These screens require less maintenance, with less frequent trimming and perhaps less watering during dry periods. I recommend using dwarf or intermediate conifer cultivars of juniper (*Juniperus*), false-cypress (*Chamaecyparis*), arborvitae (*Thuja*), spruce (*Picea*), hemlock (*Tsuga*), or fir (*Abies*). Cultivars of these conifers have attractive textures and colors with dense canopies that extend to the ground. Mixed plantings of conifers or a combination of conifers and long-lived broadleaf evergreens can be much more attractive and provide more ecological benefits in a natural setting than a single row of the same plant.

Windbreaks

Conifer screens can also serve as effective windbreaks. Of course, placement of the trees must be tailored to the location and orientation of the building and the direction of the prevailing wind. In the temperate regions of the Northwest, the prevailing winter winds are often from the northwest. If this is the case where you live, screens on the north, northwest, or west of your home can protect it from the destructive effects of high winds and reduce heat loss and thereby help conserve energy. Evergreen conifers in cold northern climates placed on the south side of a home would be quite effective in blocking southerly winter winds and lowering interior temperatures through shading in summer. However, these benefits may be offset by shading the home in the spring and fall, when solar gain is desired. Where that is the case, deciduous conifers or broadleaf trees are a better option because they can provide the summertime shading but will lose their leaves in time to clear the way for solar radiation come early spring and late fall. In warmer

Emerald green arborvitae planted in a row can serve as an effective windbreak or privacy screen.

climates, where lots of fossil-fuel energy is devoted to cooling homes, evergreen conifers (and the enduring shade they provide) are a better bet.

Conifer windbreaks should have dense foliage and be placed perpendicular to the prevailing cold winter wind. A relatively small windbreak for a home should ideally be placed no more than the expected mature conifer height upwind from the building. If the windbreak is placed too far from the building, it can create a cold sink on the leeward side of the windbreak. When designing your hedge, be sure to provide adequate space for your plants to mature. Suggested conifers for windbreaks include eastern red-cedar (*Juniperus virginiana*), western red-cedar (*Thuja plicata*), mugo pine (*Pinus mugo*), Austrian pine (*Pinus nigra*), Scots pine (*Pinus sylvestris*), Norway spruce (*Picea abies*), and Colorado spruce (*Picea pungens*) cultivars.

Accent or Focal Point

When it comes to drawing the eye, sheer beauty reigns supreme. Of course, beauty is in the eye of the beholder, so for more specific direction in determining an accent or focal point, consider shape and color. Narrow columnar or pyramidal conifers can serve as strong, dominant accents or focal points in a garden, as they contrast with or offset horizontal garden elements and underplantings of shrubs and perennials. The vertical accent serves to pull the viewer's eye through the space, which promotes a fuller three-dimensional experience of the landscape. Here are some examples:

- The tall, narrow *Cupressus nootkatensis* (synonym *Chamaecyparis nootkatensis*) 'Green Arrow' makes a striking accent with its very narrow form and downward weeping branches. *Pinus strobus* 'Angel Falls' is a refined weeping white pine with very long, light green needles.
- The relatively narrow, conical *Pinus engelmannii* 'Bush's Lace', with its downward cascading branches and attractive baby-blue needles, makes a real impact against a dark background.
- *Abies concolor* 'Blue Cloak' is a dense, upright, pyramidal white fir with attractive powder-blue foliage.
- *Cedrus libani* 'Green Prince' is a naturally bonsai-looking dwarf with rich-green needles and perfectly horizontally tiered, yet twisted branches.
- *Larix decidua* 'Krejci' is a very slow-growing larch (3 to 4 inches per year) with sparse branching and knobby clusters of buds that continue to increase in size each year. A truly unusual and unique specimen.
- *Tsuga heterophylla* 'Thorsen's Weeping' produces a striking cascading form when staked.

Color is another powerful tool when using plants as focal points. Plants that turn golden or bright yellow in winter can bring life to a mostly dormant garden or brighten an entryway. For the list of conifer cultivars that are yellow/golden in winter, but green in summer, and those that are yellow/golden year-round, along with their characteristics, see tables in the Appendix. These are some of my favorites:

- *Pinus contorta* var. *latifolia* 'Chief Joseph' has bright yellow-gold needles in winter.
- *Pinus strobus* 'Louie' is likely to attract attention in any landscape, with its year-round soft-golden needles.
- *Abies nordmanniana* 'Golden Spreader' is a choice bright gold selection; it is especially attractive in winter.

For sheer beauty, consider these:

- *Abies koreana* 'Kohout's Ice Breaker', with its two-toned shimmering silvery recurved needles; it is a real showstopper.
- The well-known *Picea orientalis* 'Skylands'. This tree has the whole package, with its perfect narrow pyramidal shape, somewhat pendulous graceful branches, and bright yellow new growth, along with bright red pollen cones.
- *Cedrus libani* 'Green Prince'. I find it a wonderful accent in my garden. Its natural upright form and beautiful horizontally tiered, yet twisted branches, with clusters of short dark-green needles tightly attached, make it unique and quite eye-catching.

Picea abies 'Anita's Golden Cloak' in the garden of conifer enthusiast David Olszyk (behind a stunning young specimen of *Abies koreana* 'Kohout's Ice Breaker'). It was selected by Bob Fincham from a cross of *Picea abies* 'Acrocona' × 'Gold Drift' and named after Dave Olszyk's wife, Anita.

Structure and Backbone

Conifers can provide a framework for the garden as a whole or act as a harmonizing element among many different plants. Integrating a conifer—perhaps a large or small pyramidal one—into an island bed or mixed border brings cohesiveness to the design and anchors it. Conifers can provide a dominant, enduring element or serve as a backdrop to vignettes of deciduous shrubs, perennials, flowering annuals, groundcovers, and shimmering grasses blooming and fading with the change in seasons. A well-placed conifer also gives a sense of naturalness and depth to planting beds and borders.

A combination of conifers, appropriately selected and placed and varying in size, form, texture, and color, can create an invigorating multilayered landscape.

Multilayered Property Border

If yours is a relatively small residential yard, the border of your property is an ideal place to blend conifers with native and site-adapted broadleaf plants, perennials, grasses, and bulbs. A multilayered community of plants mimics a natural plant community, facilitates natural processes, creates depth through the layering of planes, and provides a lush environment for nature's pest control agents. This will give your landscape structural complexity and functional diversity, and it will be inviting to a variety of wildlife species and make your garden more sustainable. Eventually, the different plant groups will grow into an interwoven plant community. Some suggestions for conifers to blend into your multilayered landscape are *Cryptomeria japonica* 'Black Dragon', *Chamaecyparis obtusa* 'Kosteri', and *Picea pungens* 'Baby Blue Eyes'.

Foreground, center, *Juniperus communis* 'Corielagan' and foreground, right, *Juniperus rigida* var. *conferta* 'Silver Mist' front a large, red Japanese maple and blend beautifully into the multilayered, interwoven landscape.

Groundcovers

Just as hedges or screens need not be made from multiples of the same plant, groundcovers can be a mixture of prostrate (spreading) low-growing plants. A mixture of spreading conifer mounds of differing colors and textures can add more vibrancy than a single prostrate conifer; plus they are low maintenance—nearly all are easy-care, drought-tolerant plants with dense growth that effectively keeps out weeds.

Historically, junipers have been the groundcover workhorses in both commercial and residential settings. However, the combination of the overuse of certain cultivars, inappropriate selection, and high flammability has given junipers a bad name. However, if they are appropriately selected they can fulfill an important function in the garden. Note that many communities in the fire-prone West are recommending the use of less-flammable plant material and/or placement of junipers well away from structures. The four juniper species that have been used extensively for groundcovers are *J. horizontalis, J. communis, J. squamata,* and *J. rigida* var. *conferta,* all of which prefer full sun.

- There are many green cultivars of *Juniperus horizontalis,* as well as several nice blue varieties, such as 'Blue Chip' and 'Blue Forest', and yellow varieties such as 'Mother Lode' and 'Golden Carpet'.

A groundcover combination of *Juniperus communis* 'Glywd' and *Juniperus horizontalis* 'Golden Carpet' surround *Cupressus glabra* 'Blue Pyramid'.

- For a tough-as-nails, very low-growing, cold-hardy variety, try *Juniperus communis* 'Corielagan'. This green juniper grows only 1 inch (2.54 cm) high and about 10 feet (3.05 m) wide, with a growth rate of 6 to 8 inches (15.24 to 20.32 cm) per year.
- Most junipers have a coarse texture, but *Juniperus rigida* var. *conferta* 'Silver Mist' has a finer texture and attractive silvery blue-green foliage in spring that turns a coppery bronze in fall. It grows to about 2 feet (0.61 m) high and 10 feet (3.05 m) wide, with a growth rate of 6 to 8 inches per year, and it is hardy to zone 6. It's excellent on slopes or cascading over a rock wall and is tolerant of salt and partial shade.
- For decorating a wall, I have used the cultivar *Juniperus rigida* var. *conferta* 'Emerald Sea'. This medium-textured green juniper grows 12 to 20 inches (30.48 to 50.80 cm) high and 6 to 10 feet (1.83 to 3.05 m) wide. The female plants with berries are particularly attractive when draped over a wall.

Juniper 'Emerald Sea' mixes beautifully with manzanita 'Pacific Mist' to elegantly drape over a rock wall.

Today, we have many new conifer groundcover cultivars suitable and available for almost any landscape application. Many of the other conifer genera that make excellent groundcovers have colors and textures that are often superior to coarse-textured juniper cultivars.

- The unusual Russian arborvitae (Microbiota decussata) has flat, fan-like, soft and feathery bright-green leaves that transform to bronze-purple in fall and winter. It is tolerant of cold temperatures and cold, dry winds and is surprisingly heat tolerant, as well. It does fine in both full sun and partial shade, growing to about 1 foot (0.30 m) tall and spreading to 12 feet (3.66 m) or more.
- *Pinus sylvestris* 'Hillside Creeper' is a vigorous intermediate prostrate conifer, ideal for covering a bank or wall.
- *Picea orientalis* 'Ferny Creek Prostrate' is a slow-growing (3 to 5 inches/7.62 to 12.70 cm per year), mounding, and spreading Oriental spruce.
- One of my favorites is the deodar, or Himalayan, cedar (*Cedrus deodara*) 'Prostrate Beauty', with its distinctive soft, feathery blue foliage; the new

needles in spring are light blue-green. In my garden, it has grown about 2 feet (0.61 m) tall and 8 feet (2.44 m) wide over 10 years, with no evidence of a leader emerging. It is in a sunny area near the top of a southwest-facing slope that is extremely dry in summer. It never receives supplemental water and is doing just fine. It is hardy only to zone 7 (10 to 0°F/–12.2 to –17.8°C), so it would not be appropriate for the upper Midwest or Northeast.

- The Colorado spruce *Picea pungens* 'Procumbens' is an attractive low-spreading conifer particularly suited for covering slopes or walls. Its undulating branches and bright blue foliage are hard to beat. It grows 6 to 8 inches (15.24 to 20.32 cm) per year to 2 feet (0.61 m) high and 10 feet (3.05 m) wide. It is hardy to zone 2 (–50 to –40°F/–45.6 to –40°C).
- An exceptional hardy and adaptable Northeast native, the jack pine cultivar *Pinus banksiana* 'Schoodic', was discovered on the Schoodic Peninsula in Maine. This emerald green, low-growing spreader tolerates cold, heat, and dry or poor soils. It grows 3 to 5 inches (7.62 to 12.70 cm) per year to 2 feet (0.61 m) high and about 4 feet (1.22 m) wide and prefers full sun.
- For shady areas, try the plum-yew *Cephalotaxus harringtonii* 'Prostrata'. It has lime-green needles in spring that transform to deep green in summer. Its

The groundcover Japanese black pine *Pinus thunbergii* 'Mt. Hood Prostrate' or 'Yatsubusa' featured in the Iseli Nursery demonstration garden.

undulating branches grow slowly, from 3 to 6 inches (7.62 to 15.4 cm) per year to 6 feet (1.83 m) wide and 3 to 4 feet (0.91 to 1.22 m) tall. It is drought tolerant and cold hardy to zone 6 (−10 to 0°F/−23.3 to −17.8°C).

We have covered most of the major uses of garden conifers, but I would like to briefly mention some other applications for conifers in the garden landscape, as the potential uses of these versatile plants are as varied as one's imagination.

Conifers as Support for Clematis

Sometimes rules can be broken for the sake of a greater good—in this case, the hydrozone rule of placing plants together with similar water and cultural requirements—for the sake of creating a beautiful garden vignette that warms the heart and soothes the soul. I am talking specifically about having spring or summer deciduous flowering clematis growing on selected conifers. Many clematis thrive in loamy, moderately moist, well-drained soil with added slow-release organic rose and flower food (4-6-2: the proportion [%] of nitrogen (N), phosphorus (P), and potassium (K) in the fertilizer). However, there are quite a few clematis that prefer lean soil, especially species from the steppe regions. Within the genus, some horticultural groups are much more adaptable than others. Large-flowered hybrids are the fussiest. Do some research before selecting and growing your clematis.

There are numerous deciduous clematis varieties well-suited for growing on pyramidal or columnar conifers. Most flowers of cultivars from the Viticella, Atragene, Tangutica/Orientalis, or Texensis clematis groups will provide a striking display because their colors complement the color of conifer foliage. However, in temperate climates with clay soil, clematis from the Atragene group requires excellent drainage; it is important that water does not pool around the crown. Clematis from the Tangutica/Orientalis group originate from the steppes; they require lean soil and annual hard pruning, and they tolerate lots of sun. The Texensis hybrids require full sun from the ground up to flower well. Following are some good choices for conifer-clambering clematis:

- A clematis with yellow flowers—such as 'Golden Harvest' of the Tangutica/Orientalis group or *Clematis otophora,* a semi-evergreen vine best in partial shade—contrasts perfectly with a conifer that has dark-green foliage. Also, from the Tangutica/Orientalis group, 'Bill Mackenzie' complements conifers with blue foliage, and 'Lambton Park' is good with blue foliage (plus, it is fragrant).
- A clematis with purple flowers, such as 'Etoile Violette' and 'Polish Spirit', both in the Viticella group, contrasts well with light green or golden foliage.

Clematis × 'Niobe' is an outstanding vine with leathery, textured green foliage and large, brilliantly colored wine-red blooms.

When you bring a new clematis home, it is best to cut back one-half of the plant's foliage without delay to reduce the demand on the root system before it is transplanted. If it has flowers, you may let it flower in the pot, but as soon as the flowers are gone, prune it back before transplanting. Although many clematis prefer their roots in the shade and their flowers in the sun, many others tolerate or prefer sun from the ground up.

There are two good approaches for planting clematis under conifers:

1. If space allows, and the conifer's canopy does not reach the ground:
    - Plant the clematis out toward the drip line (the outside edge of the conifer's bottom branches) in the ground—without a pot, if it is easy to dig a hole. Only large-flowering hybrids require that their crown be buried by 3 inches (7.62 cm) of soil or mulch; the smaller flowering plants need their crown covered by only an inch (2.54 cm) or so of soil or mulch.
    - Install a small panel of chicken wire or fencing cloth to give the clematis a "ramp" onto the conifer.
2. If you want to plant a clematis inside the drip line, closer to the trunk:
    - Plant the clematis in peat-free, well-draining soil in a whisky barrel or plastic or terracotta pot from which the bottom portion has been removed. Place this barrel or pot near the trunk. The clematis roots will find their way into the tree's soil, and you won't need to chop any tree roots.
    - Once the clematis is well established, cut the pot away. Add soil around the exposed roots to make a small berm.

Once a clematis is established, each year provide a couple of tablespoons of slow-release organic rose and flower food (4-6-2), first in March, and again whenever you prune the clematis during the growing season. Prune as necessary to keep the plant healthy, to train it to go where you want it to go (or not to go), and for the maximum floral effect possible for the species or hybrid. For more information on clematis selection, pruning, and other care, see *Plant Lovers Guide to Clematis* by Linda Beutler (Timber Press, 2016); also, take a look at the website of the Rogerson Clematis Garden www.rogersonclematiscollection.org, which has some useful handouts you can download.

Dwarf Conifers for Containers

Dwarf conifers can provide varied shapes and year-round color and texture in containers for patios, balconies, decks, and rooftops. Here are some dwarf conifer cultivars I've found particularly useful and beautiful for containers:

1. The Serbian spruce *Picea omorika* 'Peve Tijn' is a dense globose to pyramidal spruce with two-tone blue-and-green foliage that turns gold when exposed to full sun.
2. The Scots pine *Pinus sylvestris* 'Moseri' is a very dense, globose to pyramid-shaped pine with tufted needles at branch tips. The dark green becomes a brilliant amber-gold in winter.

3. The eastern white pine *Pinus strobus* 'Green Twist' is a compact, mounded pine with curly, twisted, bluish-green needles yielding a delightful, refined texture.
4. The Japanese pine *Pinus parviflora* 'Aoi', is a slender, narrow upright pine with shimmering blue color and purple cones in spring.
5. *Picea pungens* 'Blue Pearl' is a miniature spruce with unparalleled blue color and an incredibly compact, uniformly rounded shape.
6. Cedar of Lebanon *Cedrus libani* subsp. *brevifolia* 'Kenwith' with short, erect branches and small rich-green needles is an exceptional cultivar for a container.

Framing a Special Vista

Pyramidal conifers are perfect for framing a view, be it in the garden itself or beyond. Their vertical form can guide the eye to "borrowed" scenery beyond the garden and help create the illusion of increased expansiveness. They can also be used to frame your house or a garden feature, such as a fountain, large stones, or a beautiful vase.

Providing Shade to Hardscapes

The dense foliage of conifers is particularly effective for providing year-round shade on hardscapes such as driveways, walkways, and patios, thereby significantly reducing soil, ground, and ambient temperature on warm, sunny summer days.

Pines in a Japanese-Style Garden

In Japanese-style gardens, pines are treated as living sculptures, shaped through pruning off unwanted twigs and branches, removing old needles by hand, cutting or stripping some of the newly formed needles, and breaking off pine candles to modulate how the tree grows. This process takes many years, even decades, at the hands of a skilled gardener. The time spent on subtle manipulation is more than worth it; an exquisitely pruned tree takes on an artful shape, such that the clusters of needles at the ends of the branches are full of expression, while appearing perfectly natural. The juxtaposition of sculpted conifers with fine-textured Japanese maples can invoke both stability and motion.

The bones of a Japanese-style garden also include its stones: Conifers of different sizes and shapes may be paired with one another or with a thoughtfully chosen large stone. You can also place conifers and other plants along meandering paths made of steppingstones carefully chosen for their size and surface characteristics and placed to guide the visitor along twists and turns that change the view, alternately hiding and revealing parts of the garden. As in multilayered landscapes, plants are arranged in different layers in myriad shades of green to create an asymmetrical balance that appears natural and in harmony with nature.

Head Gardener emeritus Don Brooks coaches tree pruner Keith Kelly in the ancient art of Japanese black pine pruning at Seattle's beloved Kubota Garden.

Multilayered landscape at the conifer-rich Kubota Garden in Seattle.

I hope that this chapter has inspired you to envision all the possibilities for conifers in your garden and you are now eager to explore the wide range of cultivars that can help bring your ideas for your urban garden to life.

Shopping for dwarf conifers at a plant nursery.

Selecting Conifers

Chapter 5

Many conifers have adapted traits that allow them to survive in adverse climate conditions and in disturbed environments, which means that they grow naturally in a broad range of habitats. However, conifer families (like other plant families) and their genera differ in the environmental stressors they can tolerate, making them more suited to distinct regions. For example, some conifers grow successfully in conditions of low summer rainfall, while others tolerate warm and humid summers; some can handle very cold temperatures—much colder than the average conifer can survive.

A good first approach, therefore, is to select species native to your region (or their cultivars), as they are adapted to its weather and soils and will likely meet the needs of the local wildlife. For information on which conifers are native to your region, see Table 3 in the Appendix, which lists conifer species native to North America.

North America was divided into four major areas, the western (W), central (C), northeast (NE), and southeast (SE), by the American Conifer Society. (See conifersociety.org for a description of the different North American regions.)

- The western region (W) runs from the West Coast eastward to the Rocky Mountains and extends north from the Mexican border. It includes three provinces—British Columbia, Alberta, and Saskatchewan—and thirteen states: Alaska, Arizona, California, Colorado, Hawaii, Idaho, Montana, Nevada, New Mexico, Oregon, Utah, Washington, and Wyoming.
- The central region (C) extends from the Mexico–Texas border northward to Manitoba and Ontario; Ohio is the most easterly state, and North and South Dakota and Nebraska the most westerly.
- For this book, I divided the large West and Central regions in half, with the dividing line approximating where southern Oregon meets northern California. These regions are designated Northwest (NW), Southwest (SW), North Central (NC), and South Central (SC). Western regions tend to have very dry summers.
- The northeast region (NE) covers the Canadian provinces of Prince Edward Island, Nova Scotia, New Brunswick, Quebec; it also covers the area from the northern-most tip of the United States southward to Maryland, Delaware, and the District of Columbia.
- The southeast region (SE) includes Alabama, Florida, Georgia, Kentucky, Louisiana, Mississippi, North Carolina, South Carolina, Tennessee, Virginia, and West Virginia. Note: Although Virginia is part of the southeast region, in terms of growing firs, it is more like the northeast region in that firs can be grown on their own roots (that is, without grafting) in either location. However, it is difficult to grow firs on their own roots in North Carolina and southward in the southeast region. (See map of North American regions.)

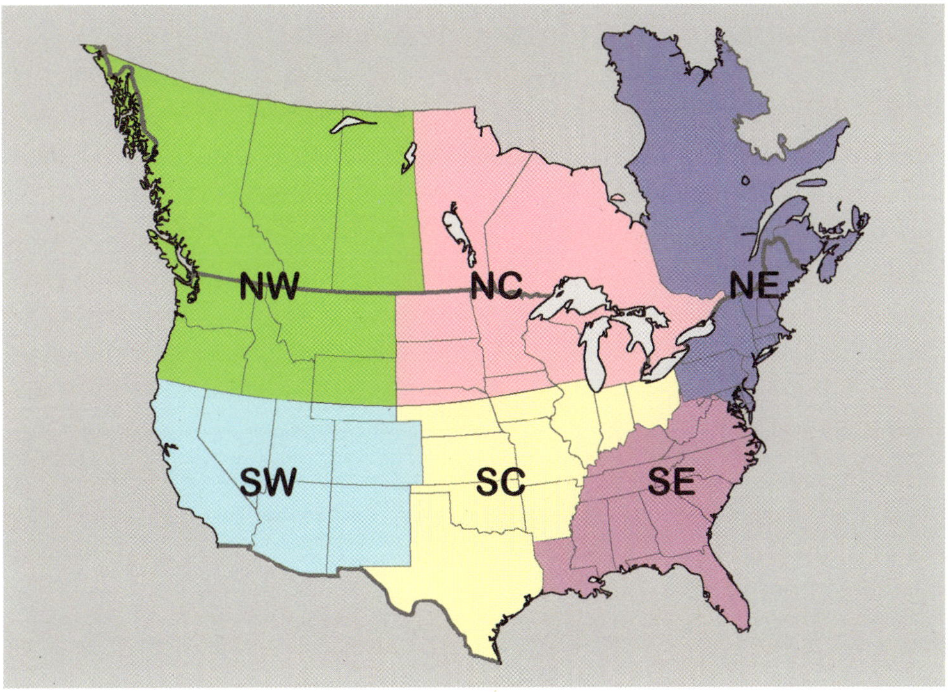

North American regions.
Credit: Mona Gilkerson

In this chapter, I provide the preferred cultural conditions for the selected cultivars, their USDA hardiness zone, and an estimate of their expected growth rate. For most cultivars, I have inferred the preferred cultural conditions and USDA hardiness zone based on information about the species as a whole. Please be aware that the preferred hardiness zone may vary depending on the specific origin or provenance of your specimen, and the hardiness of grafted conifers may be altered depending on the species chosen for the rootstock. Additionally, the growth rates generally represent averages across the United States, but the newer cultivars may only have estimates based on the specific set of circumstances where they were first grown (because it is too early to have data from other regions). It's best, therefore, to check with your local nursery for expected growth rates in your area; this is usually provided by the wholesale nursery that provided the conifer to your local nursery. Note that growth rates can vary considerably depending on the climatic and microclimatic conditions, including sun exposure, precipitation patterns, and soil conditions. The means of propagation, from a cutting or graft; the part of the plant the cutting was taken from, whether it was grafted low or high; and the species and source of understock also affect growth. Furthermore, the type of care you provide—such as watering during dry periods or fertilizing in the spring—can have a significant impact on growth rate. Thus, growth rates are only rough estimates.

I will now describe some of my favorite conifer species native to North America and their cultivars.

Conifers Native to Western North America

White fir (*Abies concolor*)

Native to the American West, *Abies concolor* 'Wintergold', is a dwarf white fir seedling selection made by Günter Horstmann of Schneverdingen, Germany, in 1959. This unique selection changes from dark green to an incredible buttery yellow with the arrival of cold weather. Then, as the weather warms in the spring, the new needles turn the tips of the yellow branches chartreuse, and the older yellow needles eventually turn dark green. It's a great cultivar to light up your winter garden. Its growth rate is about 4 to 5 inches per year. It prefers full sun to partial shade and a cool climate, as it is hardy to USDA zone 3 (−40°F to −30°F/−40°C to −34.4°C). (Note: The upper limit for most of these cultivars is not available.)

Nootka cypress (*Cupressus nootkatensis*) synonym (*Chamaecyparis nootkatensis*)

Native to North America along the West Coast across a broad geographic range, from southeast Alaska all the way down to northern California and from the high mountains down to sea level. The populations of Nootka cypress in these different regions may vary in their adaptations to local conditions, such as high elevation or the short days of Alaska winters. You should select cultivated species derived from regions that have conditions like your region.

'Green Arrow', a unique seedling selection of the Nootka cypress, was discovered on Vancouver Island, Canada, about 35 years ago and introduced to the nursery trade by Talon Buchholz of Buchholz & Buchholz Nursery of Gaston, Oregon. It was given its name because it makes a striking vertical accent, with its very narrow 2- to 6-foot-wide (0.61 to 1.83 m) form, rich blue-green foliage, and weeping branches cascading downward. It often puts out one or more ascending branches near its base. To keep its profile very narrow, you can prune off these low branches, but I prefer to keep the conifer's green skirt because it accentuates the narrow upright growth habit.

The growth rate is robust, at 1 foot (30 cm) or more per year. This Nootka cypress clone has a relatively broad physiological range, meaning it can thrive under various environmental conditions. It performs well on both the West and East Coasts and is hardy to USDA zones 4 (−30°F to −20°F/−34.4°C to −28.9°C) to 8 (10°F to 20°F/−12.2 to −6.7°C), and it generally prefers full sun, cool summers, and relatively mild winters. However, in warm climates it would benefit from afternoon shade. Although 'Green Arrow' can become quite tall in an urban environment, perhaps 30 feet (9.14 m) or more, its narrow form makes it suitable for both small and large yards.

'Sparkling Arrow' is a variegated branch selection from 'Green Arrow', discovered and introduced by Talon Buchholz. The size and growth characteristics of this narrow upright conifer are like 'Green Arrow', but it grows somewhat slower, at about 1 foot (30 cm) per year. What sets it apart from 'Green Arrow' is the dazzling creamy white variegation throughout the plant. A choice vertical accent for almost any garden.

These three distinctive, dark-green, spire-shaped Nootka cypress 'Green Arrow' cultivars stand out against the native Northwest woodland.

Colorado spruce (*Picea pungens*)

Native to the central Rocky Mountains from southern Montana and eastern Idaho south to New Mexico. Although most Colorado spruce in the wild have gray-green foliage, most of those selected for cultivation have blue foliage. Selected cultivars generally prefer moist, well-drained, slightly acidic soil and full sun, but they are tolerant of light shade; they do poorly in the heat and humidity of the South. They are generally hardy to USDA zone 3 (−40°F to −30°F/−40°C to −34.4°C).

'Blue Pearl' was derived from a witch's broom found on a *Picea pungens* 'Fat Albert' at Iseli Nursery in Boring, Oregon. This dwarf selection starts out as a dense globe with short, bright-blue needles and prominent orange-brown buds, eventually developing a broad conical habit. It grows slowly, about 2 to 3 inches (5.08 to 7.62 cm) per year. Its diminutive stature and slow growth make it particularly suitable for a small garden, rock garden, or container.

The irregular spreading and strongly drooping branches of the Colorado spruce 'The Blues' with its bright silver-blue needles forms a unique, living sculpture in the garden.

'R. H. Montgomery', a dwarf Colorado spruce seedling selection, originated in Massachusetts in the early 1930s. The plant collector R. H. Montgomery obtained it as part of his large conifer collection. It was named after him when he donated his collection to the New York Botanical Garden in 1956. 'Montgomery' is a slow-growing cultivar, about 4 to 5 inches (10.16 to 12.70 cm) per year. Its form is initially globose, but it eventually develops into a dense, broad cone with stiff, short, thick, powder-blue needles that retain their outstanding color throughout the seasons. It does fine in the Midwest, with its often warm and humid summers, but it does poorly in the heat and humidity of the South. It's a handsome accent in the shrub or perennial border or in a rock garden. Once established, after about three years, it is tolerant of extended dry periods.

Colorado spruce 'R. H. Montgomery' embodies a slow-growing, compact, broadly conical conifer with powder-blue needles.

'Ruby Teardrops' is perhaps the bluest of the dwarf Colorado spruces. What sets this cultivar apart is its production of an abundance of ruby-red cones each spring; this show-stopping color lasts for 8 weeks or more. It has a squat, irregular to globose growth habit.

'Mrs. Cesarini' was discovered as a witch's broom by Joseph Cesarini on an estate in Long Island, New York. It forms a dense, flattened mound with short, shiny blue-green needles. The profusion of light-green buds at the tips of the branches contrast nicely with the older needles. This selection usually has two flushes of growth each year, one in spring and another in summer. Its growth rate is about 3 inches (7.62 cm) per year. It will mature to about 2.5 feet (0.76 m) tall and 4 feet (1.22 m) wide in 10 years. It is, thus, ideal for a container or the rock garden.

Colorado spruce 'Ruby Teardrops' has bright blue needles and long-lasting, ruby-red cones emerging from the tips of nearly every branch in spring.

Colorado spruce 'Mrs. Cesarini' is a neat small, dense conifer for small spaces and rock gardens. Its profuse pale buds and light-green new needles contrast with the older mature needles, generating an attractive texture.

Left: *Picea pungens* 'Niemetz', aka, Spring Gold Colorado spruce is a show-stopper in spring when a creamy flush bursts open in striking contrast against last year's older blue foliage.

Lodgepole pine (*Pinus contorta* var. *latifolia*)

Native to interior Alaska and the Northwest Territories, east to Saskatchewan and the Black Hills of South Dakota, south to Colorado, central Utah, and eastern Oregon. The seedling 'Chief Joseph' was discovered by Doug Wills of Sandy, Oregon, while on an elk-hunting trip in the Wallowa Mountains of northeast Oregon; he named the cultivar after the well-known Chief Joseph, who once led the Nez Perce American Indian tribe that lives in this region. This cultivar is distinctive for its dramatic unfolding of green needles to brilliant gold as the season transitions from fall to winter. It is rather slow growing, from 4 to 6 inches (10.16 to 15.24 cm) per year, with an irregular upright habit. Like most pines, it prefers full sun and well-drained soil and is hardy to USDA zone 5 (−20°F to −10°F/−28.9°C to −23.3°C).

Lodgepole pine 'Chief Joseph', with its brilliant yellow needles, creates a show-stopping accent in winter.

Engelmann spruce (*Picea engelmannii*)

Native primarily in the higher elevations of the Rocky Mountain and Cascade Mountain ranges in western North America. 'Blue Magoo', a cultivar discovered in 1986, was introduced to the marketplace by Buchholz & Buchholz Nursery of Gaston, Oregon. It has an upright, symmetrical branching habit. The new spring growth droops somewhat, yielding a weeping effect before stiffening up in summer. The new spring needles are a striking teal blue, while the tips of the branches show off small purple cones that transform to dark brown as the weather warms. It is a great choice for very cold regions, as it is hardy to USDA zone 2 (−50°F to −40°F/−45.6°C to −40°C).

Engelmann spruce 'Blue Magoo' is an attractive upright specimen with drooping new growth and purple-red cones in spring.

'Bush's Lace' is a cultivar selected by Dick Bush of Aurora, Oregon, at a local Christmas tree farm. It has a vigorous upright form—branches extend outward before cascading downward on this relatively narrow specimen. It grows at a rate greater than 1 foot (30 cm) per year. It prefers full sun and cold climates, as it is cold hardy to USDA zone 3 (−40°F to −30°F/−30°C to −34.4°C) and may not perform well in warm climates above zone 6 (−10°F to 0°F/−23.3°C to −17.8°C), except in the Northwest, with its mild summers and cool summer nights.

The bluish-green needles of the Engelmann spruce 'Bush's Lace' contrast nicely with the dark, purple-leaf plum in the background.

Douglas-fir (*Pseudotsuga menziesii*)

The Douglas-fir has wide distribution along the Pacific coast, from British Columbia to Mexico and east to the Rocky Mountains. It is the state tree of Oregon. The cultivar 'Mt. Angel Blue' was obtained from a witch's broom found on a Douglas-fir at a golf course in Mt. Angel, Oregon, by Stephanie Krieg Pratt and was introduced by Conifer Kingdom in 2014. Unlike the species, which has green needles, this compact dwarf Douglas-fir has bright blue needles and produces a profusion of cones at the branch tips. It grows slowly, at about 3 to 5 inches (7.62 to 12.70 cm) per year and is hardy to USDA zone 5 (−20°F to −10°F/−28.9°C to 23.3°C). It's a very attractive new dwarf conifer.

Douglas-fir 'Mt. Angel Blue' derived from a witch's broom found at Evergreen Golf Course in Mt. Angel, Oregon, by Stephanie Pratt. It forms a compact blue-green tree.

Western white pine 'Halloween Broom' originates from a rare witch's broom on a western white pine tree; it was discovered by Sam Pratt on Halloween in Port Orchard, Washington.

Western white pine (*Pinus monticola*)

Native to the Sierra Nevada, Cascade, Coastal, and Rocky Mountain ranges. *Pinus monticola* 'Halloween Broom' was discovered by Sam Pratt in Port Orchard, Washington, on Halloween in 2015. It grows 2 to 3 inches (5.08 to 7.62 cm) per year and is hardy to zone 7 (0°F to 10°F/–17.8°C to –12.2°C). Its silver-blue needles and small cones make this cultivar unique—and rare—because *Pinus monticola* rarely has witch's brooms. Since this variety was introduced recently, it is still uncommon. It is unique in the Northwest because there are very few dwarf pines native to this region.

Giant sequoia (*Sequoiadendron giganteum*)

Native to the Sierra Nevada mountain range in California. 'Blauer Eichzwerg' is a dwarf seedling selection of the giant California sequoia made by Herbert Rihacek in Eichgraben, Austria, in 1989. It has an upright, dense conical habit with short, stout symmetrical branching, and prickly blue-gray foliage in summer that changes to gray-blue in winter. It is slow growing, at only about 2 to 3 inches (5.08 to 7.62 centimeters) per year and is hardy to USDA zone 6 (−10°F to 0°F/−23.3°C to −17.8°C). It likes full sun but tolerates partial shade. It is drought tolerant once established.

Dwarf giant sequoia 'Blauer Eichzwerg' is a choice dwarf selection of the giant sequoia that will easily fit into a small garden.

Conifers Native to Central North America

White spruce (*Picea glauca*)

Native to the northern temperate and boreal forest of North America. The cultivar 'Blue Teardrop' developed as a sport on the *Picea glauca* cultivar 'Echiniformis' in 1990 at Coenosium Gardens in Aurora, Oregon. It forms a dense small pyramidal tree with short, light blue-green needles. It has a growth rate of 6 inches (15.24 cm) per year and is very cold hardy to USDA zone 2 (−50°F to −40°F/−45.6°C to −40°C), preferring cold climates with cool summers. It does not do well in the eastern United States in USDA zones above 6 (−10°F to 0°F/−3.3°C to −17.8°C). It can be used for integration in a shrub border or planting bed or serve as a stand-alone specimen.

White spruce 'Blue Teardrop' is a compact, very cold-hardy conifer with short blue or silver-blue needles and dense branches.

Bald-cypress (*Taxodium distichum*)

A long-lived, highly adaptable deciduous conifer native to the bayous, swamps, and rivers of the Southeastern and Gulf coastal areas and the lower Mississippi River valley. It forms a strong central leader with symmetrical branching, creating a pyramidal silhouette 50 to 70 feet (15.24 to 21.34 m) at maturity. Although it typically grows in swampy water in the South, it does very well on a wide range of soils, including drier sites and colder areas in the North, as it is hardy to USDA zone 4 (−30°F to −20°F/−34.4°C to −28.9°C). It has fine-textured yellow-green leaves that emerge in spring, changing to a stunning orange-cinnamon in the fall before dropping off. The reddish-brown exfoliating bark adds appeal. This is a great specimen plant for full sun in larger gardens and parks. For small gardens, there are several excellent cultivars available. I like the compact weeping cultivars.

'Cascade Falls,' is a seedling selection found on a lake bank in Auckland, New Zealand, and introduced into the United States by Larry Stanley of Stanley and Sons Nursery of Boring, Oregon. If this cultivar were not grafted, it would simply sprawl on the ground; it needs to be staked to a desired height to allow the branches to weep down. Alternatively, 'Cascade Falls' can be grafted onto *Taxodium distichum* understock at the desired height to avoid the need for staking. This cultivar is hardy to USDA zone 5 (−20°F to −10°F/−28.9°C to −23.3°C).

'Peve Minaret' is another great cultivar of the bald-cypress; it was selected by Piet Vergeldt of the Netherlands. It grows slowly to about 6 feet (1.83 m) tall and 2 to 3 feet (0.70 to 0.91 m) wide in ten years. Its general form mimics that of the species, but the foliage at the ends of the branches is tighter and smaller, appearing like tiny minarets. It is hardy to USDA zone 5 (−20°F to −10°F/−28.9°C to −23.3°C).

Bald-cypress 'Cascade Falls', seen here in the author's garden, has soft, yellowish-green new foliage in spring and orange-cinnamon fall foliage prior to leaf drop. It's an exceptional, highly adaptable, deciduous conifer.

Conifers Native to Eastern North America

Eastern white pine (*Pinus strobus*)

A handsome five-needle eastern white pine native to northeastern United States, southeastern Canada, and parts of the upper Midwest of the United States. Its form is initially pyramidal, but it matures into a broad oval habit with an irregular crown. It has long, soft, bluish-green needles. This white pine is susceptible to blister rust. Cool, moist habitats favor the disease. The first symptom is yellow or red spots on infected needles. Use only plant-resistant stock, and don't plant near nonresistant currents or gooseberries, the principal alternate hosts of the blister rust fungus.

The cultivar 'Green Twist' comes from a witch's broom found in 1996 on the *Pinus strobus* cultivar 'Torulosa' in the Heartland Collection at the Bickelhaupt Arboretum in Clinton, Iowa. It forms a dense globe with twisted branches bearing lovely blue-green, finely textured curling needles. It grows slowly, about 2 inches (5.08 cm) per year, to

Eastern white pine 'Green Twist' is a fine-textured, dwarf conifer with ornamental twisted blue-green needles.

about 5 feet (1.52 meters) tall and 3 feet (0.91 meters) wide. It's hardy to USDA zone 4 (–30°F to –20°F/–34.4°C to –28.9°C). Grow it in full sun and well-drained acidic soil and protect it from strong winds. 'Green Twist' is moderately drought tolerant, but water it regularly during dry periods, especially during the first few years. It is tolerant of urban pollution and resists pests and disease. This unique dwarf conifer is perfect for a container or planter box or near a patio or terrace where it can be viewed regularly.

'Louie' is a moderately fast-growing cultivar with attractive soft-golden outer needles that stay vibrantly colored throughout the year; its internal foliage stays light green. It was introduced by Greg Williams of Kate Brook Nursery, who found it as a seedling in Hyde Park, Vermont, in the 1990s. It forms an open, upright, symmetrical pyramid and grows approximately 1 foot (30 cm) per year, but its growth can be considerably curtailed by pruning the new growth (candles) each spring and lightly shearing the new foliage. As the needles do not scorch, it is best to grow this tree in full sun. It should light up any landscape.

Sometimes it takes experimentation to decide on your conifer's permanent home. Planting small conifers in a container can allow you to try out different places before final planting. Here, eastern white pine 'Louie' looks stunning in its temporary container in late afternoon light in David Perry's garden.

'Beal's Starry Night' is named after Professor William James Beal of Michigan State University. It is a rare cultivar recently found as a witch's broom on a *Pinus strobus* planted by Professor Beal in 1896 as part of a forestry experiment. This slow-growing (2 to 3 inches or 5.08 to 7.62 cm per year) conifer has tufts of very short light green needles at the branch tips that look like stars around each bud.

Pinus × *schwerinii* 'Wiethorst' is a witch's broom selection from a hybrid between the Himalayan pine (*Pinus wallichiana*) and the eastern white pine (*Pinus strobus*), found by Johann Wieting of Giesselhorst in the German town of Schwerin. This pyramidal pine produces a profusion of large, handsome, resinous cones and long, fine blue-green needles. It is hardy to USDA zone 4 (−30°F to −20°F/−34.4°C to −28.9°C). This delicate-looking hybrid makes a dramatic background or impressive accent.

Because many conifers can adapt to different environmental conditions, we have the luxury of growing quite a few conifers from different regions of North America. Furthermore, North Americans can grow conifers from many regions of the world. Good online sites to view photos and descriptions of available conifers include www.conifersociety.org, www.coniferkingdom.com, www.brokenarrownursery.com, and www.westernevergreen.com/home.html. To further aid you in your conifer selection, the

Pinus × *schwerinii* 'Wiethorst', with its abundance of large, resinous cones and long super-fine, bluish-green needles, creates an outstanding accent in the landscape.

North America regions where many commonly grown conifer species and their cultivars can be grown are shown in table 5.1.

Table 5.1: North American Regions where Conifer Species and their Cultivars Can Be Grown

Title	NW	SW	NC	SC	NE	SE
Abies alba	Y		Y		Y	Y
Abies balsamea	Y		Y		Y	
Abies concolor	Y	Y	Y		Y	
Abies koreana	Y		Y		Y	Y
Abies lasiocarpa	Y	Y	Y		Y	
Abies nordmanniana	Y		Y		Y	Y
Abies pinsapo	Y	Y	Y	Y	Y	Y
Abies procera	Y				Y	
Cedrus atlantica	Y	Y	Y	Y	Y	Y
Cedrus deodara	Y			Y	Y	Y
Cedrus libani	Y	Y	Y	Y	Y	Y
Chamaecyparis lawsoniana	Y		Y		Y	
Chamaecyparis obtusa	Y		Y		Y	Y
Chamaecyparis pisifera	Y		Y	Y	Y	Y
Cryptomeria japonica	Y	Y	Y	Y	Y	Y
Cupressus nootkatensis	Y		Y	Y	Y	Y
Larix decidua	Y		Y		Y	
Larix kaempferi	Y	Y	Y		Y	Y
Larix laricina	Y		Y		Y	
Metasequoia glyptostroboides	Y		Y	Y	Y	Y
Picea abies	Y	Y	Y	Y	Y	Y
Picea engelmannii	Y	Y	Y		Y	Y
Picea glauca	Y	Y	Y		Y	Y
Picea mariana	Y		Y		Y	Y
Picea omorika	Y	Y	Y	Y	Y	Y
Picea orientalis	Y		Y	Y	Y	Y
Picea pungens	Y	Y	Y	Y	Y	Y
Pinus banksiana	Y		Y		Y	Y
Pinus cembra	Y		Y		Y	
Pinus contorta	Y	Y	Y		Y	
Pinus densiflora	Y	Y	Y	Y	Y	Y
Pinus flexilis	Y	Y	Y	Y	Y	Y
Pinus heldreichii	Y	Y	Y	Y	Y	Y

Table 5.1: North American Regions where Conifer Species and their Cultivars Can Be Grown

Title	NW	SW	NC	SC	NE	SE
Pinus koraiensis	Y	Y	Y		Y	Y
Pinus mugo	Y	Y	Y	Y	Y	
Pinus nigra	Y	Y			Y	
Pinus parviflora	Y		Y		Y	Y
Pinus pumila	Y		Y		Y	
Pinus resinosa	Y		Y		Y	
Pinus strobus	Y		Y	Y	Y	Y
Pinus sylvestris	Y	Y	Y	Y	Y	
Pinus thunbergii	Y	Y		Y	Y	Y
Pinus wallichiana	Y		Y		Y	Y
Pseudotsuga menziesii	Y	Y	Y		Y	
Sciadopitys verticillata	Y		Y		Y	Y
Taxodium distichum	Y		Y	Y	Y	Y
Taxus baccata	Y		Y		Y	Y
Thuja occidentalis	Y		Y	Y	Y	
Thuja plicata	Y		Y		Y	
Thujopsis dolabrata	Y		Y	Y	Y	Y
Tsuga canadensis	Y		Y		Y	
Tsuga heterophylla	Y		Y		Y	
Tsuga mertensiana	Y	Y			Y	

As each of these areas is rather large and they vary considerably in their hardiness zones and specific microclimates, this table is a starting point to give you a reasonable idea of which species and their cultivars may survive and grow well in your region. However, it's imperative that you match your conifer to your specific planting conditions, as discussed in detail in Chapter 3. Notice that all the species listed can do well in the Northwest region. Note that *Cryptomeria japonica* can be grown in all six North American regions but only in hardiness zones 5 or 6 or higher, depending on the specific cultivar. So, because large parts of the North Central region are colder than zone 5, this species would not be suitable for all North Central regions. Similarly, the very adaptable *Taxodium distichum* can grow in all regions. However, although it can tolerate the dry summer conditions of the NW, it is not likely to do well in most of the SW without considerable supplemental water because of extreme drought in much of this region.

Conifers Native to Regions Outside of North America

Korean fir (*Abies koreana*)

A compact conical evergreen native to the mountains of Korea. The branches are covered with short needles that are shiny and dark-green on top and silver underneath. Grow it in moist, slightly acidic, well-drained soil in full sun or part shade. It's hardy to USDA zone 5 (−20°F to −10°F/−28.9°C to −23.3°C), but it will not do well in heavy clay or hot and humid summer conditions; it prefers a cool summer climate. As it is not particularly drought tolerant, you'll need to provide adequate water during extended dry periods. There are many wonderful Korean fir cultivars available in North America today. One of my favorites is 'Kohout's Ice Breaker', a dwarf cultivar derived from a witch's broom found on the well-known, relatively slow-growing cultivar 'Horstmann's Silberlocke'. It was found by the German plantsman Jörg Kohout in his display garden. The needles are particularly strongly curled and twisted around the branches, prominently displaying the shimmering silvery portion of the needles. The radial branches are short and stubby. Consequently, the cultivar has a small globose habit but eventually develops a leader and forms a small, squat tree. Once the plant is established, its expected growth rate is about 2 to 3 inches (5.08 to 7.62 cm) per year.

An outstanding specimen of the Korean fir 'Kohout's Ice Breaker', with the Caucasian spruce 'Bergman's Gem' in the background.

Himalayan cedar (*Cedrus deodara*)

This species has a broad pyramidal habit when young, with the branches gracefully pendulous. The cultivar 'Prostrate Beauty' is a prostrate, flat, horizontally spreading selection discovered by Iseli Nursery of Boring, Oregon, in 1979. Its densely layered branches are covered with attractive soft, feathery, distinctive-looking light blue-green needles. It grows about 6 to 8 inches (15.24 to 20.32 cm) per year and is hardy to USDA zone 7 (0°F to 10°F/−17.8°C to −12.2°C), performing best in full sun and well-draining soil. It is a unique and eye-catching drought-tolerant groundcover, perfect for cascading over walls, covering slopes, or highlighting a patch of level ground.

The new, powder-blue foliage of a narrow, upright deodar cedar 'Bush's Electra' blends beautifully with the leaves of a deciduous eastern redbud tree.

Cedar of Lebanon (*Cedrus libani*)

Native to Lebanon, western Turkey, and south-central Turkey, this stately tree has a massive truck, very wide-spreading horizontal branches, and a flat-topped crown. Its natural climate is one of warm, dry summers and cool, moist winters. It requires full sun and excellent drainage. The cultivar 'Green Prince' was discovered in 1969 as a dwarf seedling by Neil Hall of Wells Nursery in Mount Vernon, Washington, among a group of *Cedrus libani* trees in Oregon. He took cuttings, then propagated them and introduced the conifer as 'Minuta Wells Green Prince'. It has rich, dense, short dark-green needles and beautiful horizontally tiered, twisted branches. New needle growth is light green, which contrasts nicely with the older dark-green needles. It prefers somewhat dry soil, so good drainage is a must. It is hardy to USDA zone 5 (−20°F to −10°F/−28.9°C to −23.3°C) and very drought tolerant. I have two specimens in my garden. They have never needed watering, even during the very dry Northwest summers. When I purchased this cultivar in 1985, I was told that it was a miniature conifer with a growth rate of less than ½ inch per year. However, both of my 'Green Prince' specimens—the second one I purchased in 2007 from Wells Nursery—are growing approximately 4 to 5 inches (10.16 to 12.70 cm) per year. This growth rate still qualifies it as a dwarf, and it is a good illustration of how growth rate can vary depending on the circumstances. This exquisite dwarf cedar cultivar, with its stout form and slow growth, makes a great centerpiece or focal point for a small landscape.

The elegant cedar of Lebanon 'Green Prince' in the author's garden in late afternoon winter light.

The cute small, dense mound of cedar of Lebanon 'Hedgehog' with its long, prickly needles resembles a spiny hedgehog.

Cedrus libani 'Hedgehog' is a seedling selection of cedar of Lebanon made by Cedar Lodge Nursery in New Plymouth, New Zealand, in 2009. This dwarf cultivar creates a low, tight-growing mound with layered horizontal branches bearing long, densely packed, rich blue-green prickly needles. It looks kind of like a cute and colorful hedgehog. It grows about 1 inch (2.54 cm) in height per year and perhaps closer to 2 inches (5.08 cm) in width. Grow it in well-draining acidic soil in full sun.

Serbian spruce (*Picea omorika*)

A narrow pyramidal tree with pendent branches that tend to rise at the tips, native to southeastern Europe. It has glossy dark-green leaves and grows slowly to about 50 feet (15.24 m) in cultivation. It is tolerant of a wide range of soil conditions but prefers moist, well-drained soil. It is cold hardy to USDA zone 3 (−40°F to −30°F/−40°C to −34.4°C), performing well in the Northwest, upper Midwest, and Northeast; it does poorly in the hot and humid Southeast. It is a wonderful specimen for blocking unwanted views or as an accent in a larger garden.

'Peve Tijn' originated from a golden sport on the *Picea omorika* cultivar 'Nana', found by Piet Vergeldt at his nursery in Lottum, the Netherlands, and named after his son. This dwarf cultivar is globe-shaped when young and transforms into a small conical tree as it ages. It is lushly clothed in two-toned light blue-green needles with tips highlighted a pleasant yellow. The combination of the three colors creates a stunning color combination. It grows slowly, from 3 to 6 inches (7.62 to 15.24 cm) per year, and it is hardy to USDA zone 3 (−40 to −30°F/−40°C to −34.4°C). It tolerates air pollution but not hot and humid summers. While it is best to plant it in full sun in moist, well-drained soil, it will do fine in light shade. An unusual dwarf cultivar for the container, planter box, or patio edge.

This impressive Serbian spruce (below, right), with its drooping leader brings a unique sense of character to the author's wintery garden. To see and compare it to a more typically slender, upright example of *Picea omorika* (Serbian spruce), turn to page 169.

'Pendula Bruns' originated as a seedling selection at the Bruns Nursery in Germany. This is a relatively fast-growing (6 to 12 inches [15.24 to 30.48 cm] per year), upright, narrow, strictly weeping cultivar (ends of the branches come straight down), with needles that are blue-green on top and silvery on the undersides. It would be a dramatic focal point for any garden.

Caucasian spruce (*Picea orientalis*)

Native to the north-shaded slopes of the Caucasus Mountains in southern Russia and northeast Turkey, the Caucasian spruce forms a compact narrow pyramid with horizontal branches that tend to be somewhat pendulous. Bright red pollen cones and short, shiny, dark-green needles appear on top of the branches in spring. It is hardy to USDA zone 4 (−30°F to −20°F/−34.4°C to −28.9°C). The cultivar 'Skylands' was introduced by Skylands Botanical Gardens of New Jersey. It has bright yellow, new growth in spring and usually holds its color well in cool climates, but it may fade in the heat of summer. To maximize its color, plant in full sun, although be aware that some needle burn may occur on young plants. It is adapted to a wide range of soils, though it grows best—about 1 foot (0.30 m) per year—in well-drained soils with medium moisture and some protection from winter winds. This attractive, graceful cultivar does well in the Northwest, Midwest, and Northeast but not in the heat and humidity of the South.

Caucasian spruce 'Skylands' with a profusion of bright-red pollen cones.

Japanese white pine (*Pinus parviflora*)

Native to the subalpine regions of Japan and Korea, this small conifer has a dense conical habit when young and stiff; its often-twisted needles form five-needle tufts at the end of the branches, giving the tree an attractive fine texture. Hardy to USDA zone 4 (–30°F to –20°F/–43.4°C to –28.9°C), it does well in full sun and moist, well-draining soil, and it is tolerant of salt spray. The cultivar 'Ōgon janome' or, "Eye of the Golden Snake," has distinct yellow banding on the blue-green needles near their base. This strikingly beautiful pine grows about 6 to 10 inches (12.7 to 24.4 cm) per year.

'Tanima no yuki' or, "Snow in the Valley," starts out as an irregular mound but transforms into a small, broad, upright pine. Creamy white needles emerge from light pink candles in spring. It grows about 3 to 4 inches per year and may benefit from some partial shade in the afternoon. It would look great in a container.

Japanese white pine 'Tanima no yuki' (Snow in the Valley). Inset, its seed and pollen cones.

Scots pine (*Pinus sylvestris*)

Widely distributed from northern Europe to northern Asia and Siberia, and naturalized in many other regions of the world, this fast-growing, medium-sized conical tree has distinctive scaly red-brown bark and twisted blue-green needles in bundles of two. It is hardy to USDA zone 3 (−40°F to −30°F/−40°C to −34.4°C) and prefers moist, acidic, well-drained soil in full sun. It is tolerant of urban pollution, but it does not like the heat and humidity of southern summers. The dwarf cultivar 'Green Penguin' seedling selection was found by Jim Lewis of J Farms in Amity, Oregon, in the late 1990s while he was working at a nursery in Park Rapids, Minnesota. This upright, bushy pine, with dense, thick, rich-green needles, produces tufted short juvenile needles in late summer that surround the longer mature needles. Its shape makes it easy to imagine why this pine conjures up the appearance of a penguin. This unusual dwarf conifer grows slowly, from 2 to 5 inches (5.08 to 12.70 cm) per year. It would be an excellent choice for a container or integrated into a shrub border.

The distinctive Scots pine 'Green Penguin' is squat, like its namesake.

Pinus sylvestris 'Moseri' is another dwarf cultivar of Scots pine. This seedling was selected by Moser Nursery in Versailles, France, in 1900. The needles have two flushes of growth, with long, dense needles forming in spring and short, dense needles in fall. Surprisingly, these green needles turn a brilliant golden yellow in winter. This cultivar grows slowly, at about 2 to 5 inches (5.08 to 12.70 cm) per year. It is hardy to USDA zone 3 (−40°F to −30°F/−40°C to −34.4°C). Grow it in full sun. This cultivar is an excellent choice to brighten up your winter garden.

Chinese red pine (*Pinus tabuliformis*)

Also known as table-top pine, this northern China and Korea native naturally develops a flat top as it matures. It grows to between 30 and 50 feet (9.1 to 15.2 m) in cultivation, though its seedling can also make an excellent bonsai specimen due to its attractive horizontally spreading branches. It is hardy to USDA zone 5 (−20°F to −10°F/−28.9°C to −23.3°C). The cultivar 'Jiuzhaigou Valley' was found as a witch's broom by Jörg Kohout in Jiuzhaigou Valley National Park in the northern Sichuan Province in southwest China. It is named after this park and nature preserve. *Pinus tabuliformis* is very uncommon in cultivation, and 'Jiuzhaigou Valley' is one of the very few cultivars of this species. It grows into a broad, globose shrub. The long thin needles and smooth purplish stems on spring's new growth are quite ornamental.

Chinese red pine 'Jiuzhaigou Valley'.

Golden-larch (*Pseudolarix amabilis*)

Although this is commonly called golden-larch, it is not a true larch. This deciduous conifer native to China was introduced into cultivation by plant explorer Robert Fortune in 1854. In cultivation, it grows relatively slowly, to between 30 and 50 feet (9.1 and 15.2 m) tall, with a spread of 20 to 40 feet (6.1 to 12.2 m). It has a very open conical shape, with wide-spreading, elegantly layered horizontal branching. Beautiful, soft, long, flattened, bright-green needles with two glaucous stomatal bands on the undersides appear in clusters on short shoots and singly on long shoots. In fall, the leaves turn a golden brown. It is hardy to USDA zone 5 (−20°F to −10°F/−28.9°C to −23.3°C) and best grown in full sun in moist, acidic, well-drained soil. A most elegant specimen as a focal point or accent for a larger garden.

Selecting Conifers for Specific Conditions: Right Plant, Right Place

Most conifers do best in sunny conditions and slightly acidic, reasonably moist, non-salty soil with good, but not excessive, drainage. However, you can find conifers that break this mold; some are able to tolerate and even thrive in more adverse conditions. This strength has allowed conifers to spread far and wide across the globe.

Urban soils often differ from native soils in that they can have poor drainage, contain various pollutants (including road salts), and may have an alkaline pH. Thus, they often provide less-than-ideal conditions. Even so, it is likely you will find one or more conifer species adapted to your specific urban conditions. Let's consider conifer selections for a variety of potentially difficult situations or adverse conditions.

Wet Soil

Poorly draining soil that is periodically flooded during the growing season tends to remain wet for long periods. The long wet period makes the soil oxygen deficient. A variety of factors cause boggy soils; the most common are a high water table, excess clay, compaction, and naturally low-lying areas.

Plant roots normally absorb oxygen and expel carbon dioxide in the process of respiration. Those plants not adapted to wet soils can suffer potentially fatal root suffocation from lack of oxygen and/or root disease from living in an overly wet environment. In addition, trees and shrubs experiencing decline from excessive water are more susceptible to secondary diseases and harmful insects. However, as is the case with almost every part of an ecosystem, there are pros and cons, depending on where you stand. Wet areas can indeed offer some ecological benefits, such as removing contaminants from water, helping to recharge the groundwater, and providing important habitat for wildlife. So, wet soil has its place, and there are plenty of conifers that naturally occur on wet sites.

- Bald-cypress (*Taxodium distichum*) commonly grows in swamps and along waterways in the Southeast of the United States. It can withstand prolonged flooding and even saltwater intrusions.
- Dawn redwood (*Metasequoia glyptostroboides*) and the American larch (*Larix laricina*) are deciduous conifers especially tolerant of some flooding.
- Several other conifers don't mind wet feet and may tolerate intermittent flooding, though not prolonged inundation. These include western red-cedar (*Thuja plicata*), Oriental arborvitae (*Pladycladus orientalis*), and eastern arborvitae (*Thuja occidentalis*).
- Swamp-cedar (*Chamaecyparis thyoides*) is native to freshwater swamps, bogs, and wet woods along the Atlantic coast.
- Black spruce or swamp spruce (*Picea mariana*) is found in swamps in the South.
- White spruce (*Picea glauca*) is commonly found alongside streams and lakes.
- Balsam fir (*Abies balsamea*) is native to cool, moist woods and bottomlands.

Dry Soil

Semiarid or steppe regions of the country—such as the Great Plains, parts of Southern California, and the Great Basin—are dry because they have low precipitation rates combined with high evaporation rates. Typical dry soils are sandy and fast draining and have low moisture holding capacity.

Quite a few conifers have evolved in parts of the world that typically have very dry soils. Examples of conifers that grow and survive on dry and often warm sites include the hard pines that generally have two or three needles per fascicle, along with spruces, junipers, cypresses, and true cedars.

- Hard pines such as Jack pine (*Pinus banksiana*), Bosnian pine (*Pinus heidreichii*), ponderosa pine (*Pinus ponderosa*), shore pine (*Pinus contorta* var. *contorta*), mugo pine (*Pinus mugo*), Austrian pine (*Pinus nigra*), Scots pine (*Pinus sylvestris*), pitch pine (*Pinus rigida*), Virginia pine (*Pinus virginiana*) and the single-leaf piñón pine (*Pinus monophylla*) do well in dry soils.
- Spruces for dry soils include Colorado spruce (*Picea pungens*), Norway spruce (*Picea abies*), Engelmann spruce (*Picea engelmannii*), and white spruce (*Picea glauca*).
- Examples of cypresses and junipers that can be used on dry soils (in tree, shrub, or groundcover form, depending on the species or cultivar)—include Arizona cypress (*Cupressus arizonica*), Italian cypress (*Cupressus sempervirens*), Monterey cypress (*Cupressus macrocarpa*), common juniper (*Juniperus communis*), creeping juniper (*Juniperus horizontalis*), shore juniper (*J. rigida* var. *conferta*), and eastern red-cedar (*Juniperus virginiana*).
- The true cedars, Atlas cedar (*Cedrus atlantica*), deodar cedar (*Cedrus deodara*), and cedar of Lebanon (*Cedrus libani*), are particularly tolerant of hot and dry conditions. Although not a true cedar, the incense-cedar (*Calocedrus decurrens*) is also tolerant of hot and dry conditions.

Finally, and perhaps somewhat unexpectedly, I have found that the groundcover Russian arborvitae (*Microbiota decussata*) can withstand the very dry Northwest summers. In my garden, it does fine on a southwest hillside in full sun.

Alkaline Soils with High pH

Most conifers prefer slightly -to -moderately acidic soil with the pH ranging between 5.5 and 6.5. Conifers planted in alkaline soils with a pH above 7 are likely to suffer from nutrient deficiency as several minerals, such as manganese and iron, become limited. However, the optimal range for each individual species can vary considerably, and there are some genera or their species that can tolerate alkaline soils. These include the following:

- Arborvitae (*Thuja*)
- Juniper (*Juniperus*)
- Cedar (*Cedrus*)
- False-cypress (*Chamaecyparis*)
- Cypress (*Cupressus*)

- Several spruce species, including *Picea glauca, Picea omorika*, and *Picea pungens*
- Several pine species, including *Pinus nigra, Pinus leucodermis*, and *Pinus mugo*

Hot and Humid Areas

These species do well in challenging regions for most conifers, such as the southeastern United States:

- Momi fir (*Abies firma*)
- Monkey puzzle tree (*Araucaria araucana*)
- Incense-cedar (*Calocedrus decurrens*)
- Cedar of Lebanon (*Cedrus libani* subsp. *libani*)
- Japanese plum-yew (*Cephalotaxus harringtonii*)
- Hinoki-cypress (*Chamaecyparis obtusa*)
- Sawara-cypress (*Chamaecyparis pisifera*)
- Swamp-cedar (*Chamaecyparis thyoides*)
- Chinese-fir (*Cunninghamia lanceolate*)
- Arizona cypress (*Cupressus arizonica*)
- Chinese juniper (*Juniperus chinensis*)
- Shore juniper (*J. rigida* var. *conferta*)
- Eastern red-cedar (*Juniperus virginiana*)
- Dawn redwood (*Metasequoia glyptostroboides*)
- Asian bayberry (*Nageia nagi*)
- Lacebark pine (*Pinus bungeana*)
- Japanese red pine (*Pinus densiflora*)
- Slash pine (*Pinus elliottii*)
- Serbian spruce (*Picea omorika*)
- Caucasian spruce (*Picea orientalis*)
- Loblolly pine (*Pinus taeda*)
- Japanese black pine (*Pinus thunbergii*)
- Oriental arborvitae (*Platycladus orientalis*)
- Buddhist-pine (*Podocarpus macrophyllus*)
- Bald-cypress (*Taxodium distichum*)

Areas with Light Shade or Dappled Shade

Urban gardens are at the whim of not only the sun's arc but also the surroundings and, in a growing city like Seattle or Bellevue, Washington, ever-changing landscapes. In an already sun-scarce region, shade amplifies plants' struggle to photosynthesize and grow. It is therefore important to match each conifer with the amount of light available to it. Conifers that prefer light shade or dappled shade have evolved to be specifically adapted to low-light conditions; consequently, they have a high photosynthetic efficiency, which means they are able to convert a relatively high proportion of light energy into chemical energy during photosynthesis. Some conifers that meet these criteria include

- Japanese plum-yew (*Cephalotaxus harringtonii*)
- Japanese yew (*Taxus cuspidata*)
- English yew (*Taxus baccata*)
- Canadian yew (*Taxus canadensis*)
- Hinoki-cypress (*Chamaecyparis obtusa*)
- Hiba arborvitae (*Thujopsis dolabrata*)
- Hemlock species, such as western hemlock (*Tsuga heterophylla*), Canadian hemlock (*Tsuga canadensis*), and northern Japanese hemlock (*Tsuga diversifolia*)

Conifers that do best in partial shade or dappled shade but do fine in full sun if they get enough water include

- Eastern arborvitae (*Thuja occidentalis*)
- Western red-cedar (*Thuja plicata*)
- Balsam fir (*Abies balsamea*)

Other conifer species (and even more so their smaller cultivars) that can acclimate and do well in relatively low-light situations, particularly if they have a few hours of morning sun, include

- Japanese-cedar (*Cryptomeria japonica*)
- Caucasian spruce (*Picea orientalis*)
- Alberta or white spruce (*Picea glauca*)
- Norway spruce (*Picea abies*)
- Nootka cypress (*Cupressus nookatensis*)
- Japanese umbrella-pine (*Sciadopitys verticillata*)

- Korean fir (*Abies koreana*)
- Min fir (*Abies recurvata*)
- Japanese white pine (*Pinus parviflora*)
- Eastern white pine (*Pinus strobus*)
- The conifer groundcover from the cypress family Russian arborvitae (*Microbiota decussata*) is also tolerant of shady conditions.

Conifers that normally prefer full sun (six or more hours of direct sunlight per day) may do just as well or better in partial sun, with no more than four hours of direct sunlight per day, if grown in warm climates in the South or areas of the mild West Coast that have dry summers.

Windy, Exposed Areas

The tug and pull of strong winds can alter conifers' root–soil relationships, cause breakage of shoots, and contribute to shorter growth and abnormal development. Winds also increase water loss through evaporation. To minimize water loss on windy sites, provide adequate water to trees and shrubs and surround them with a 4-inch-thick layer of coarse mulch or wood chips.

Conifers with small, narrow leaves tend to be wind resistant. Many pines, junipers, and spruces are quite wind resistant. These include

- Jack pine (*Pinus banksiana*), mugo pine (*Pinus mugo*), Scots pine (*Pinus sylvestris*), shore pine (*Pinus contorta* var. *contorta*), and Austrian pine (*Pinus nigra*)
- Rocky Mountain juniper (*Juniperus scopulorum*), Savin juniper (*Juniperus sabina*), eastern red-cedar (*Juniperus virginiana*), and Chinese juniper (*Juniperus chinensis*)
- White spruce (*Picea glauca*), Colorado spruce (*Picea pungens*), and Sitka spruce (*Picea sitchensis*)

Other wind-tolerant species are bald-cypress (*Taxodium distichum*), dawn redwood (*Metasequoia glyptostroboides*), Port Orford-cedar (*Chamaecyparis lawsoniana*), Irish yew (*Taxus baccata* 'Stricta'), and eastern arborvitae (*Thuja occidentalis*).

The next chapter will guide you through the options for obtaining the conifers you have in mind and getting them off to a healthy start for a long life in your garden.

Planting Conifers

Chapter 6

As conifers are usually quite expensive, make sure you match conifers to your specific site and microclimate and plant them properly. Select a site with healthy, fertile soil, good drainage, good air circulation, and an appropriate amount of sunshine or shade for the selected species.

Planting a conifer at the right time is key to planting success. Ideally, plant when the conifer is dormant, from late fall or mid-November to late winter or mid-March. The optimal time will depend on your specific climate conditions. Select a time when the top 10 inches of soil is moist, the temperature is cool (between 32° and 60°F), and the soil is not frozen solid. Thus, in areas with dry summers and mild winters, such as the West Coast west of the Cascade Range, it is best to plant in the late fall or winter when rainfall is abundant and the temperatures are cool but not bitterly cold. In the northern parts of the country, where the soils are frozen solid during winter, a more optimal time to plant conifers may be in late summer/early fall or in early spring when the soil has thawed and a hard frost is not likely to occur. It's best to plant in the early morning on an overcast or cloudy day to reduce water loss from evaporation or transpiration.

How to Plant: Balled and Burlapped (B&B)

Before you purchase a balled and burlapped or container-grown conifer, you'll want to check it for good health. Look for signs of insect damage or exterior browning on the foliage (interior browning of some conifers is normal and does not pose a problem). To examine the roots of a containerized conifer, place the container on its side, tap the container, and gently slip the plant out to examine it, looking for the following attributes:

- Is the peripheral root zone developed sufficiently to hold soil in the root ball?
- Is the peripheral root zone overdeveloped to the extent that relatively large roots are circling the root ball?
- Do the roots appear healthy? Look for white or cream-colored root tips, which suggest active growth.
- Are the roots waterlogged or discolored? Is there evidence of the root cortex (outer covering) sloughing off?
- Are parts of the roots missing or look as though they have been chewed by insects? The roots should be evenly distributed from the top to the bottom of the container.
- Are there root defects such as circling, hooked, or kinked roots, or is the main root "J" hooked? Root-bound plants and those with defects should be avoided.

B&B trees are offered by nurseries primarily for trees with trunks 2 inches or more in diameter. They are field-grown in native soil. As they are harvested, most of the roots are removed, and the remaining roots and soil are wrapped in burlap.

Assess your site, assess your soil, purchase your conifer. Now comes the fun part.

If planted according to the following instructions, most B&B trees and shrubs have high transplant survival rates.

1. Make sure the roots in the entire root ball are moist. If in doubt, soak the B&B roots in water for 20 to 30 minutes. Keep the roots moist and shaded until they are planted.
2. Set the root ball or network of roots and associated soil next to the proposed planting spot. Hold and transport the tree by the root ball, not by the trunk. Identify the root flare—the area of transition between the trunk and the roots where the trunk widens and connects to the roots (if it is covered by soil, expose it)—and measure the distance between the root flare and the bottom of the root ball.
3. Dig a saucer-shaped hole with sloped sides, with the bottom dug only1 inch shallower than the depth of the root ball. The width of the top of the hole should be about three times the root ball's diameter and the bottom about two times the diameter. Planting trees too deep is a common mistake that must be avoided. You should always avoid disturbing the soil at the bottom of the hole because you do not want the plant to settle too deep. If you find you have dug the hole too deep, add soil back to the bottom and firm it to raise the surface. Do not amend the planting hole with compost.
4. Carefully center the root ball in the hole. The root flare should be slightly (1 to 2 inches) above the top surface of the undisturbed ground surface around the hole. It is best to have a long, flat object like a yardstick or 2×4 handy so you can check the correct ground level.
5. Pull the burlap back and away from the root ball. Although not truly necessary unless the burlap is treated and nondegradable, I prefer to cut away the material with scissors. It's best to leave the burlap on the bottom of the root ball to minimize the likelihood of root ball disturbance. In arid regions and soils with low moisture, remove as much burlap as possible, as it will not decompose well in arid regions. If there is a wire basket, cut it back as low as possible.
6. Inspect the surface of the root ball for damaged, kinked, or hooked roots, and remove them with a clean, sharp pruner.
7. Backfill the hole to about half full by packing the original native soil around the root ball and pressing to remove all air pockets.
8. Water thoroughly and let the water in the hole drain, then completely backfill with the remaining loose soil, making sure that the trunk flare is completely exposed and slightly above grade.

9. Use any extra soil to build a small berm (4 to 6 inches high and wide), forming a watering basin around the perimeter of the planting area. Water thoroughly and slowly to saturate the soil and remove any residual air pockets.
10. Spread 3 inches of protective mulch around the tree, keeping the mulch several inches away from the root flare and trunk.
11. B&B trees usually do not have trouble standing upright on their own. If yours seems to need support, drive two sturdy stakes into the ground on opposites sides of the tree. Tie the stakes to the tree with separate flexible ties to prevent windthrow, allowing the tree several inches of movement in all directions. The tree can be staked like this for up to a year. Remove the ties as soon as the tree seems well-rooted. Staking can reduce the normal taper of the trunk, making the tree weaker and more likely to snap in the wind.
12. Keep the soil and mulch around your young tree moist. During dry weather, water the tree deeply down below the roots to the drip line (outermost circumference of the conifer canopy) every seven to ten days for at least a year, until it is established.

The downside of this planting approach is that it does not initially allow much contact between the roots in the root ball and the native soil. Also, differences in soil texture between the root ball and the native soil in the planting hole can interfere with water flow at the interface of the different soils. Consequently, the roots are more likely to dry out even if regularly irrigated during dry periods. The most likely scenario with B&B trees is for the root ball to have a heavy clay texture and the native soil to have a sandier loam with good drainage. If this is the case, the water will have a difficult time moving into the root ball from the surrounding native soil during the first growing season; the water tends to run off the heavy clay soil in the root ball, leaving most of the roots to dry out. This often leads to the death of the tree. Therefore, when planting B&B trees grown in clay, it is important to make sure that the root ball is completely moist after watering. One solution is to remove the outer portion of the soil from the root ball before planting to expose the feeder roots; this ensures that the feeder roots are in direct contact with the native soil. You would then follow the procedure for bare-root planting of container-produced trees. However, I recommend taking this approach only when the conifer is dormant—to better ensure that it can withstand the shock of bare-root planting.

How to Plant: Container Grown

Container-grown plants are generally much lighter than B&B plants and, therefore, easier to handle and transport. They are commonly grown in a plastic container in a "soil-less," porous, fast-draining synthetic planting mix. The major advantage is that the root system is developed in the container, which allows all the roots to be moved when transplanted, thereby significantly reducing transplant shock. The disadvantage is that the plant roots can reach the bottom and sides of the pot, start spiraling and, in a relatively short time, become pot-bound and tangled. Fortunately, a few commercial growers are beginning to adopt innovative alternatives to the commonly used nursery pot system. One innovative new container design guides the roots into outward-projecting cones in which the roots reach the open air and die back, stimulating the development of a dense, healthy mass of fibrous roots near the root tips, which help the plant establish itself more quickly after transplantation. Thus, this kind of container eliminates spiraling roots and their associated problems at planting time. Hopefully, this kind of container innovation will be adopted by more conifer growers and nurseries.

This dwarf mugo pine which had been growing to size in a container has had its rootball untangled and is ready to be transferred to a permanent home in the ground.

Conifers grown in containers can potentially be successfully planted throughout the growing season, but to minimize transplant shock and enhance the likelihood of success, I recommend planting in late fall or early spring, when the conifer is dormant and the soil is not frozen hard. Prior to planting, water the containerized tree regularly to keep the soil mix damp; check it daily during warm periods to make sure the soil mix does not dry out. Plant on a cool, cloudy morning to minimize water loss and the potential for the roots to dry out.

1. To ensure that the roots are moistened and to facilitate removal of root-bound plants from the container, soak the container with the plant in water for 20 to 30 minutes, then drain the excess water—but keep the roots moist while transplanting.
2. Tip the moistened container onto its side and tap the sides and bottom to loosen the root system from the pot until the conifer can slide out. Rather than planting the "soil-less" fast-draining planting mix along with the plant roots, gently bare the roots by removing as much of the container mix as is practical from between the roots. This helps minimize interface problems between the container mix and the

native soil (which often contains clay), giving all the feeder roots direct contact with the native soil in the planting hole. Note that some conifers, such as the giant sequoia (*Sequoiadendron giganteum*) and its cultivars, have very fragile feeder roots that break off easily. The process of removing soil mix or native soil will break the feeder roots. Furthermore, this conifer is quite sensitive to dehydration. Therefore, it is important not to "bare root" this conifer.

3. Dig a saucer-shaped hole with sloped sides almost as deep as the root mass and three to four times the diameter of the root spread at the top of the hole and two to three times the root spread at the bottom of the hole. If the soil is high in clay, leave a firm central mound of undisturbed soil at the bottom. (This mound is not recommended for sandy loam because of the potential for settling, which would leave the root flare too deep.) With or without the mound, the hole should be deep enough to accommodate all the roots without crowding or bending them. When using the mound, adjust the height of the hole bottom (or soil mound) so the plant's root flare will be slightly above the surrounding grade, with the roots spreading outward and downward around the sides of the mound.
4. Untangle and tease apart the roots with your hands, and, using a sharp pruner, remove those portions of roots that are damaged, circling, kinked, or hooked. It's best to determine if the plant is extensively root-bound (with circling roots on the bottom and sides) at the plant nursery, so you can leave it there and find a better specimen.
5. If you're providing a central mound in the hole, spread the roots radially and downward over the mound while holding the tree in an upright position, adjusting as necessary so that the trunk flare is slightly above the surrounding grade.
6. Still holding the plant at the correct depth, backfill with loose, moist native soil until the hole is about three-quarters full, firming the soil around the roots to make sure all are in direct contact with native soil, with no air pockets.
7. Water the plant thoroughly and let the water drain; then backfill until the hole is filled.
8. Build a small berm (3 to 4 inches tall and wide) around the perimeter of the planting hole. Water thoroughly and deeply to moisten all of the backfill material and surrounding soil.
9. If the young tree cannot stand on its own, support it by tying it to a stake on each side. The prevailing wind should be parallel to the line demarcated by the two poles and the tree. Tie loose and low (about one-third of the tree

height from the ground) with flexible ties and leave in place until the tree can stand on its own—which should be no longer than one year.

10. Remember to water the conifer regularly during dry periods until well established in the garden, every seven to ten days—or more frequently, depending on soil moisture. It is possible to overwater conifers, especially if the roots are not yet actively growing. White pines and firs are prone to root rot as they are getting established if given too much water or are not given adequate drainage.
11. Control weeds and other competing vegetation around the conifer for at least two years to give it time to get established without competition.

Sometimes you don't have a choice, and you must plant a containerized conifer when it is not dormant. In this case, follow the previous directions, but with one primary difference: Keep the soil mix intact after you remove the plant from its container. If you like, you can gently loosen and remove some of the soil mix just at the surface of the outside feeder roots. Also, you should scarify the sides of the hole with a shovel, and the conifer should be soaked in the container for 30 minutes prior to planting. To minimize interface problems, make sure the planting site has excellent drainage. If you have clay and/or compacted soil, well before planting day, you should mix 3 inches of fine compost into 12 inches of soil throughout the planting area, not only in the anticipated planting hole.

How to Plant: Bare Root

Except for young seedlings, conifers are generally not sold bare root because of the high potential for root desiccation. It is best to plant a bare-root conifer seedling immediately, but if necessary you can store it for one or two days at just above freezing (34°F to 38°F—you can use your refrigerator on a low setting, if necessary), with the roots kept moist and the top shaded.

1. Before planting, prune back any roots that are not firm and plump, then soak them in a bucket of water overnight.
2. Prepare the planting hole as described for container plants, with a solid soil cone in the center of the hole to support the root system during planting.
3. Spread the roots over the firm soil cone and backfill with the soil from the hole. Gently compact the soil with your hand while making sure that the root flare is at the level of the surrounding ground.
4. Water thoroughly. If the seedling settles below its intended position, raise it slightly to the preferred position and firm the soil to remove any air pockets.
5. If necessary, stake the seedling and tie it loose and low, allowing it to move in the wind. Water as necessary to keep the roots moist.

How to Plant: Containers

Containerized evergreens can provide color and texture around your house, apartment, or condo throughout the year—a welcome brightening element, particularly in winter. Miniature and dwarf conifers are ideal for containers because they grow slowly, retain their leaves year-round, and require little maintenance. Consider making trough gardens on your deck, porch, or balcony (or in the garden) by planting these mini trees in containers of stone, hypertufa, or other lightweight material.

Because conifers grown in containers are exposed to greater temperature variations than conifers planted in the ground, be sure to select compact, slow-growing, winter-hardy cultivars. In general, conifer roots are more sensitive to temperature extremes than are the trunk or branches, so choose conifers hardy to temperatures considerably colder

Dwarf and miniature conifers are perfect for small gardens or container gardens. Their unique year-round features really shine in decorative containers, which can easily be moved around. Pictured is a dwarf mugo pine.

than your natural climate. If you live in USDA hardiness zone 6, for example, choose a plant that is hardy to two zones lower, or zone 4. In the case of grafted trees, a different, more hardy species could be used as a rootstock to increase the conifer's hardiness. Regardless, container conifers may need some winter protection. Your goal should be to reduce temperature fluctuations. Adding mulch around the pot up to near the rim will help. If possible, move the container to a protected location away from cold winds and out of full sun. If you live in a very cold climate, you can sink a container in the ground in the fall before a hard freeze or wrap insulation material around the pot.

When choosing a conifer cultivar, before you purchase it, make sure the specimen is not excessively root-bound and/or showing extensive circling roots. The container you plant in must be large enough to provide ample space for growing roots, and it needs to have adequate drainage.

Garden soils are usually not suitable for plants in containers because they are heavy and tend to compact over time. Instead, use a good commercial soil mix made specifically for growing plants in containers; this mix should provide both adequate moisture and nutrient retention as well as fast drainage. Before planting, add water to the bag of soil mix to thoroughly and evenly moisten it; leave some excess water in the bag overnight to make sure all the soil is moist. Poke holes in the bottom of the soil bag to allow any excess water to drain before placing the soil in the container.

1. Soak the new conifer for 30 minutes prior to planting in the container.
2. Fill the pot with the moist potting mix up to 1 inch below the container rim, and lightly tap down the soil.
3. Remove enough of the soil mix from the center of the pot to allow enough room for the plant, reserving the mix.
4. Make sure the conifer's current container is drained of excess water, then tip it onto its side, tap the sides and bottom, and gently remove the plant, keeping the soil mix around the roots intact.
5. Place the conifer's roots, soil attached, in the center of the new pot so the root flare is just below the top of the pot (where the soil level will be once planted).
6. Backfill the container with the commercial soil mix and firm around the roots to a level approximately 1 inch below the top of the pot while holding the plant with its root flare in the correct position.
7. Water the soil regularly to keep the soil from drying out—but not so wet that the roots could rot.

Container plants require watering more frequently than in-ground plantings. They also require small amounts of nutrients in the form of slow-release fertilizer during growth in early spring. Be aware, however, that providing too much fertilizer stimulates excessive soft, new growth, which encourages attacks by insects and pathogens.

For examples of selected conifer cultivars for containers, see "Dwarf Conifers for Containers," in Chapter 4.

Transplanting

There are a variety of reasons why you might want to move a conifer. It may be getting too much shade; it may be competing with other plants for space and nutrients; or you may observe or anticipate it outgrowing its current location. To transplant successfully, you need to move both the root ball and as much of the feeder roots and taproot as possible. It's best to do this in the late fall, when the conifer is dormant. In colder regions, it may be necessary to transplant in late winter or early spring to avoid digging and planting when the ground is frozen solid. Tree growth is significantly reduced in dormancy, which allows the roots to better focus on getting established in the soil before above-ground growth and demand for water begins in the spring. The following technique is relatively easy with young, small conifers:

1. When the soil is slightly moist, but not too wet or waterlogged, prepare the hole in the new location that approximates the expected root ball.
2. Dig gently around the plant's existing roots to determine the location of the feeder roots and taproot, if present. Then with a clean, sharp spade, make a circular cut all around the plant, taking care not to damage the feeder roots or taproot.
3. Dig just under the root ball without cutting any feeder roots and gently loosen it from the surrounding native soil. For conifers with tap roots, try to gently remove as much taproot as possible.
4. Place burlap or a tarp next to the hole and gently remove the conifer from the hole by lifting it up by the root ball.
5. Place the root ball on the burlap and tightly wrap the soil ball and roots with burlap.
6. Keep the root ball intact while moving the root ball to its new location, and put the plant in the new hole such that the root flare is slightly above the level of the ground and its orientation to the sun is the same as in its original location.
7. Backfill with native soil from the planting hole, and fill with water to remove air bubbles and saturate the soil. Let the water drain and then rewater the planting area.

For larger conifers, the safest approach is to prune the roots ahead of time—approximately 6 to 12 months in advance. This allows the conifer to develop a flush of new feeder roots, which will reduce transplant shock. I prefer to root prune in the fall, when top growth is minimal, then transplant the following fall. You can also transplant in spring, though you will need to make sure your developing feeder roots get extra water during dry weather. It is also possible to root prune in spring and transplant in the fall.

There are several approaches to root pruning. The first, which works well with smaller conifers, involves these steps:

1. Six to 12 months prior to transplanting, cut through the outside edge of the existing roots with a clean, sharp spade, making a circular cut around the plant.
2. Transplant the root-pruned conifer with new feeder roots, as described previously.

The second approach, ideal for larger conifers, is as follows:

1. Six to 12 months prior to transplanting, when the soil is moist but not too wet, dig a trench about 1 foot wide and at least 1 foot deep around the tree's drip line.
2. Refill the trench with soil rich in organic material, ideally a mix of two parts native soil and one-part fine bark or compost. Water the trench thoroughly, especially in dry periods. The use of this trench is to encourage the feeder roots to grow into the looser soil medium.
3. After giving the new feeder roots 6 to 12 months to grow, prepare the hole that approximates the size of the root ball in the new location.
4. Carefully dig around the drip line of the conifer to be transplanted without damaging the new feeder roots. Dig under the roots and gently loosen them from the surrounding native soil.
5. Spread burlap on the ground nearby the conifer and gently move the conifer onto the burlap.
6. Tightly wrap the root ball with burlap and gently remove it from the hole.
7. Carry or carefully drag the conifer to the prepared hole. Make any necessary adjustments to the planting hole and plant the conifer so that its root flare is slightly above the level of the ground and its orientation is the same as it was in its original location.
8. Backfill with native soil and water thoroughly.

Note that conifers with significant taproots, such as many pines, some firs, and cypresses, are more difficult to transplant. It is best to transplant these conifers when they are dormant to minimize transplant shock.

Japanese black pine, *Pinus thunbergii* in a heavy, February snowstorm. (See this same iconic specimen pictured on a spring like day in March featured as the Chapter 4 opener, page 72.)

The next chapter reviews the different approaches to propagating conifers.

Propagating Conifers

Chapter 7

Propagation Terms

Cutting: a piece of stem, leaf, or root used for vegetative (asexual) reproduction. The cutting is placed in a suitable medium, such as moist sand, and, given the right conditions, a new plant will grow. Plants developed from cuttings have identical characteristics and are called *clones*.

Scion: a detached living shoot or bud from one plant that is joined to a rootstock of another to form a new plant. Typically, the top portion of the grafted plant is the scion, and the lower portion is the rootstock.

Cambium: the thin layer of tissue between the inner bark (phloem) and the wood (xylem). It is responsible for secondary growth that gives rise to the differentiation of the phloem and xylem and the healing of the scion to the rootstock.

Rootstock: a stem with a well-developed root system onto which a cutting or scion from another plant is grafted.

Root cortex: the layer just under the epidermis of a plant's root.

Bare-root seedlings: a young plant developed from seed with exposed roots free of soil.

Scarification: the process of altering or weakening a seed coat to make it more permeable to gas and water, thereby facilitating germination; for example, soaking the seed in hot water, treating the seed with acid, or scratching the seed using an abrasive surface.

Stratification: the process of simulating conditions of nature that allows seeds to break dormancy so germination can occur; for example, storing moistened seeds in a refrigerator at around 34°F (1°C) for one to three months. Different seeds have different requirements.

Plant propagation is the process of creating new plants, either sexually or asexually. You can propagate conifers from seeds—the sexual way—or with cuttings or by grafting—the vegetative, or asexual, way. There are a few reasons to choose one method over the other. If you have a beloved conifer you want to replicate, then using cuttings or grafting is your best bet, as the new conifer will have the same characteristics as its parent. With grafted plants, however, the roots will have the characteristics of the rootstock. For example, a rootstock more tolerant of heavy clay soil allows you to grow a less-tolerant grafted plant in a site with heavy soil.

If you are looking to save money and effort, or to generate large numbers in order to select new varieties relatively easily, nature's original method of seed propagation is the better way to go. But just as it is in nature, what you get is something of a crapshoot, as half the genes of the seed come from the female parent plant and the other half from the male parent. Under the right conditions, the seed matures and germinates, eventually creating a plant with some characteristics of both parents. Typically, the progeny will exhibit little variation and will resemble the species. Some seedlings, however, may have unique characteristics that can enhance their merit in the landscape.

Following are some strategies for propagating plants used in some plant nurseries. But there is no reason why you can't try these on your own. (See the sidebar for propagation and planting terms).

Seeds

It goes without saying that animals and plants need one another to survive. In a never-ending cycle of life and death, blossoming and decaying, the plants that become animal food return as plant food in the form of excrement and the animal body itself after death. Beyond this exchange of nutrients, birds, mammals, and insects are also essential actors in the process of plant reproduction. Conifers generally rely on the wind for pollination; however, five conifer families, including junipers and yews, have developed fleshy, succulent, and colorful seed coverings to attract animals. A bird swallows this tasty morsel whole and flies away, digesting the seed covering and altering it in en route. The seed itself remains undigested and intact, and, if excreted in an ideal spot, it will germinate and a new conifer will grow. Tony Holthuijzen, a biologist who studied cedar waxwings, found that when cedar waxwings pass seeds of the eastern red-cedar (*Juniperus virginiana*), the seeds are three times more likely to germinate than those not eaten by the waxwings.

Another scenario involves an animal taking the seed to a hiding place to save it for later, then forgetting about it. If the seeds are placed in a favorable spot for germination, a new plant will grow. For example, Clark's nutcracker (*Nucifraga columbiana*) feeds almost exclusively on the seeds of the whitebark pine (*Pinus albicaulis*). During the summer, it uses its long, pointed bill to tear off the cone scales and pry the seeds out; then it carries the seeds in its specialized throat pouch to seed caches nearby or to feed hungry chicks in the nest. Over the fall and winter, the bird will return to the cache; whatever seeds it overlooks can germinate in spring. Consequently, the nutcracker is largely responsible for whitebark pine seed dispersal over an ever-widening range. The Eurasian nutcracker (*Nucifraga caryocatactes*) provides the same seed dispersal service for the Arolla pine (*Pinus cembra*) in the European Alps.

Many human seed collectors rely on squirrels to first pick the seeds from cones. The seeds are then gathered from caches and processed for nursery use. You can participate in this process of propagation (eating the seed is not required). You might be called upon to mimic nature's complicated systems—depending on the species, some conifers have a complex dormancy that must be unlocked for the seeds to sprout. Dormancy ensures that the timing of germination or emergence of the seedling occurs when conditions for growth and survival of the seedling are optimal. Thus, it prevents the seeds from germinating in late fall or winter when the winter temperatures could kill the seedling. Since the type of dormancy varies with the type of seed, it is helpful to research what is needed to facilitate germination for each plant species. Some seeds have external, or seed-coat, dormancy that prevents water and oxygen from entering the seed. To break dormancy requires a treatment called *scarification*, which involves scratching the thick seed coat or helping to dissolve it through chemical means. Many conifer seeds have internal, or physiological, dormancy caused by conditions or inhibitors in the seed that prevent germination. This type of dormancy can be broken by a seed treatment called *stratification*, which typically involves soaking and chilling the seed prior to sowing to simulate the conditions in nature. Consequently, cold stratification in a refrigerator at 35–40°F (2–4°C) under moist conditions for two to six weeks, depending on the species, usually does the trick. Juniper seeds have both internal and seed-coat dormancy. Therefore, soaking the juniper seeds in weak acid before stratification increases germination.

The key step to seed propagation is to collect your seeds from the seed-bearing cones when they are ripe, or mature. Here's how:

1. Collect the cone when it is just beginning to open, which occurs when the usual green color changes to brown or gray, normally in early fall.
2. Store the cone in a warm, dry place to allow the cone to fully open. Seeds of some conifers, such as yews, junipers, and podocarpus, are found in berry-like structures, with the seed covered by fleshy scales that blend together and look like a berry. Yews and podocarpus cones are usually bright red when ripe; juniper "berries" turn a purplish black.
3. Once the cone is open, you may need to shake it to release the seed.
4. Rub or squeeze away any fleshy scale from the seed and let dry.
5. Plant seed from northern climes in the fall, when it is cool. You can plant in flats or pots or directly in the garden.
6. Cover the seed with a layer of soil that approximates the seed's diameter.
7. Water as necessary to keep the soil moist, and cover with plastic or other material, such as a heavy row cover, to help retain moisture and keep squirrels and other animals from running off with the seed.

Hopefully, by next spring some of your seeds will germinate, but don't give up hope if no seedlings sprout; some seeds may take a year or more to germinate. Once they do, let the seedlings grow for a couple years before transplanting.

As the seedlings mature, their characteristics are likely to be very similar to the tree that produced them, but variations in genetic makeup may cause them to differ somewhat. If you wish to have a plant with attributes identical to your selected plant, it is best to propagate it vegetatively by either cuttings or grafting.

Cuttings

Many, though not all, conifers can be propagated by taking a cutting from a branch and inducing it to form roots, then planting it in the ground. Arborvitae (*Thuja*), false-cypress (*Chamaecyparis*), and low-growing juniper (*Juniperus*) have been shown to root relatively easy. Cuttings of yew (*Taxus*) also root reasonably well, but hemlocks (*Tsuga*),

To propagate a yew, take a cutting roughly 8 inches from the tip of a ¼-inch-diameter branch with healthy foliage but no reproductive structures. Yew cuttings do best when collected and planted in late autumn.

pines (*Pinus*), and spruce (*Picea*) have proven more difficult. Furthermore, most dwarf conifers obtained from witch's brooms cannot be successfully propagated by cuttings. They need to be propagated by grafting. However, many dwarf hemlocks are rooted from cuttings.

The degree of difficulty in rooting varies depending on the species, the condition of the cutting source, and the time of year. Propagation is often done with semi-hardwood cuttings from the current season's growth—taken just after a flush of growth, when the wood is neither actively growing nor fully matured or dormant; the dormancy period begins in September and early October, as with false-cypress, arborvitae, yew, and Japanese-cedar. You may have better success if you take cuttings from young plants with juvenile foliage. However, from many species—such as juniper, true cedars, and hemlock—you can collect fully dormant hardwood cuttings in December. Consult Dirr and Heuser's *Reference Manual of Woody Plant Propagation* cited in the Bibliography and Resources for the best type of cutting for a given species. A general approach to prepare your cutting for rooting is as follows:

1. Prepare a rooting medium. It should be sterile, low in nutrients, and well-drained but with good moisture retention. A 50/50 mixture of coarse sand and perlite or 20/80 mixture of peat and perlite are good choices.
2. Place the medium in a clean container deep enough to allow several inches of root development below the base of your cuttings. The container must have drainage holes.
3. Place a small amount of the rooting hormone compound (powder, gel, or solution) recommended for the species in a separate container (to avoid contamination of your hormone stock.) Rooting hormone is readily available online and at nurseries and garden stores.
4. Cut 4 to 8 inches of softwood, semi-hardwood, or hardwood, cutting—as recommended for the species—from the tip of either a terminal or a lateral shoot from a healthy, insect- and disease-free conifer.
5. Remove the lower one-third to one-half of the leaves from the cutting.
6. Use a clean dowel or pencil slightly larger in diameter than the cutting to poke small planting holes in the medium, 1 to 2 inches deep.
7. Using a sharp knife sterilized with isopropyl alcohol, wound the bottom, or basal end, by making one or two 1-inch vertical cuts on the sides. For large-diameter cuttings, you can make several additional vertical cuts across the bottom of the cutting. This wounding promotes cell division and facilitates uptake of water and the root-promoting chemicals you will provide it with.

8. Dip the bottom inch of the cutting into the rooting hormone and leave it in place for a few seconds. If you are using hormone powder, once you remove the cutting, gently tap it to get rid of any excess powder.
9. Stick the bottom 1 to 2 inches of the cutting vertically into the moist rooting medium; the bottom of the cutting should be above the bottom one-third of the medium.
10. Firm the medium around the base of each cutting to ensure good contact; then add water to remove any residual air pockets.
11. Cover the container with glass or plastic to maintain humidity. Slightly vent the glass or plastic and place the container in indirect light. You may have better success rooting hardwood cuttings if you use a heat mat to keep the medium temperature between 60°F to 70°F (15.5°C to 21.1°C) until rooting has begun. You can test for rooting by gently tugging on the cutting in the growing medium.
12. Water the cuttings as necessary to keep them moist until they have rooted. The best way to ensure that the cuttings have adequate moisture is to mist them intermittently. To avoid excess humidity, make sure there is adequate ventilation in the room. Cuttings that don't survive in the medium should be removed immediately.
13. As rooting begins to take place, start hardening off the cuttings (getting them accustomed to the outdoors) by providing them more ventilation to reduce the humidity and exposing them to more sunlight.
14. Once a substantial number of roots have formed, transfer the seedlings into a container with good-quality potting soil.

Grafting

Grafting is a form of vegetative propagation in which the bottom portion (roots and stems) of a compatible plant—the rootstock—is joined with the vegetative portion of a desired plant—the scion—to create a new plant. Although several different grafting techniques can be used on herbaceous or woody plants, the type that works well on conifers is called the *side-veneer graft,* whereby the scion is joined to a dormant, potted rootstock to create a new plant. A good rootstock is adapted to the soil and climate where the grafted conifer is to be planted; it must also be compatible with the scion.

Many conifer cultivars, particularly those derived from a witch's broom (see sidebar), can be readily propagated *only* by grafting. For species that are not entirely compatible with your conditions, grafted plants can often improve a plant's growth rate and development, provided the rootstock has been shown to grow well in your microclimate conditions. You can purchase grafted plants, but before you do, make sure that the graft wound has completely healed and that there is a tight, healthy connection between the rootstock and scion. It may be best to keep some of the remaining rootstock on for a year or more, as long as it is not overgrowing the graft. When the graft appears to be growing well, carefully trim off any remaining branches of the rootstock from the graft.

A grafted plant showing the congested growth of the Douglas-fir cultivar 'Squirrel Nest' (left), while still retaining branches of the Douglas-fir rootstock (right).

If you'd like to take on the grafting process, the first step in grafting is to acquire appropriately sized potted rootstock or healthy bare-root seedlings of a species compatible with the scion. In general, the rootstock and scion must have a close botanical relationship. Obviously, you can expect reasonable success if the rootstock and cutting are from the same species, but sometimes same-species rootstock is too expensive or unavailable. Luckily, there are many examples of successful grafting with rootstock and scion that are not the same species but have a strong biological affinity. However, the longevity of the pairing may be negatively affected since the cultural requirements of the rootstock and the scion may vary significantly. Here are some examples of rootstock–scion pairings:

- The five-needle eastern white pine (*Pinus strobus*) rootstock is commonly used with many other five-needle pines.
- The southwestern white pine (*Pinus strobiformis*) is usually used as rootstock for five-needle pines in the western United States because it is better suited to the western climate.
- The Chinese white pine (*Pinus armandii*) is the predominant rootstock for white pine in Europe, but it is not yet available in the United States.
- The Korean pine (*Pinus koraiensis*) also makes a satisfactory rootstock for white pines, but it is slower growing, so it is not as economical to use.
- Good success has been obtained using the two-needle Scots pine (*Pinus sylvestris*) as rootstock for two-needle pines, three-needle pines, and all pines that are not tolerant of wet soils, regardless of how many needles are in each fascicle.
- Norway spruce (*Picea abies*) is the best rootstock for spruces because it is highly adaptable to a range of soil types. Colorado spruce (*Picea pungens*) has been used extensively as well.

Witch's Brooms

"Witch's broom" is a term used in botany for an abnormal cluster of growth from a single point on an otherwise normal plant. These deformities, which occur on conifers as well as on broadleaf trees and shrubs, can be a reaction to a variety of invading organisms, pathogens, or parasites, including dwarf mistletoes, fungi, phytoplasmas or mycoplasma-like organisms, viruses, nematodes, mites, and insects. Some such deformities have no known cause, but adverse environmental conditions may play a role. One or more of these plant stressors cause a hormonal, metabolic, or genetic alteration in a single bud or shoot. A proliferation of shoots with short internodes grow close together to form a dense mass. These shoots are usually shorter and stockier than those of the parent tree or shrub, and they can also be very different in shape, texture, and color. Their dense growth may sometimes resemble the end of a broom, but more often they resemble a bird's nest, and in fact, they can provide excellent cover and nesting sites for birds and small mammals. Witch's brooms caused by genetic mutations are usually stable and can be propagated vegetatively as dwarf cultivars. However, conifer cultivars derived from a witch's broom can be readily propagated only by grafting.

Be aware that the best rootstock and scion combination may vary with your specific climate and cultural conditions. For example, if you live in the southern portion of the United States and plan on growing a fir, it may be preferable to use Momi fir (*Abies firma*), which tolerates hot and humid conditions and heavy clay, making it a more appropriate rootstock than a northern fir, such as white fir (*Abies concolor*) or Fraser fir (*Abies fraseri*).

If possible, the rootstock should be resistant to conifer diseases common in your region. For example, the Lawson-cypress is particularly susceptible to *Phytophthora lateralis*, an aggressive soil-borne fungus-like pathogen. Use of rootstock resistant to Phytophthora (Oregon State University has developed a good one) should improve your chances of having a healthy Lawson-cypress rootstock–scion combination.

Because new information on rootstock–scion compatibility is continuously generated, check with a recently published propagation manual to provide recommendations for the best scion–rootstock combination.

How to Prepare

Always purchase more rootstock than you intend to use to allow you to match the diameter of the rootstock most closely to the diameter of your cutting or scion. Several weeks before grafting, place the rootstocks in a heated area or greenhouse to stimulate root growth. Keep the rootstock medium moist and, as cleanliness is important for successful grafting, keep the work area and the rootstock containers clean. When the roots exhibit new white tips, the rootstock is ready to graft. Discard any unhealthy-looking rootstock before you begin.

It is best to collect scions when the temperature is above freezing in December or early January, when the wood is dormant. (If it is below freezing, you'll want to slowly warm the scion by placing it in cold water or a refrigerator after cutting.) Cut the scion from a healthy terminal branch of the previous season's growth. For especially slow-growing varieties, using a cutting consisting of two or even three years' growth can make grafting easier, as thicker scions are more likely to successfully graft. Make sure that the cutting used is moist and soft enough to be easily handled during grafting. Obtain the scion wood just before the time you plan to do the grafting. If necessary, you may temporarily store the scions several days by wrapping them in moist paper towels, placing the moist scions in plastic bags, and storing them in a refrigerator between 33°F and 39°F to keep the scions dormant. Generally, success of grafting decreases significantly after scions have been cut for more than one or two weeks.

How to Graft

You'll need the following:

- A very sharp knife cleaned and disinfected with either isopropyl alcohol or Lysol. (Note: If you don't have a grafting knife, a good-quality razor blade knife will work just fine.)
- Rubber grafting tie.
- Grafting wax or beeswax.

1. Clean the lower portion of the scion by removing any dirt or needles.
2. Use a very sharp knife that has been cleaned and disinfected with Lysol to remove any needles or branches on the lower one-third of the scion.
3. Make your first cut on a smooth straight section of the rootstock, 2 to 3 inches above the root flare. In one smooth motion, make a straight downward cut at an angle into the rootstock to form a small flap on the stem, with a width that approximately matches the width of your scion.
4. Make a downward cut on each side of the scion near its base. Trim the cut surfaces so that they will approximately match the length of the cut surface of the rootstock.
5. Align the cut edges of the rootstock with the cut edges of the scion and close the rootstock flap onto the scion to eliminate air gaps and exposure of inner wood. If the cuts on the rootstock fail to match up well to those on the scion, cut a new scion.
6. Firmly tie the scion into the rootstock flap with the rubber grafting tie, wrapping from below the bottom of the cut to above the top of the cut. Apply grafting wax or beeswax to the wrapped union to seal the graft.

Pinus mugo 'Fishhook' grafted onto *Pinus mugo* rootstock with all but a small section of the rootstock removed.

Graft Aftercare

Because fresh grafts require high humidity, cover the grafted plants with clear plastic or place in a plastic case. It will also help to warm the roots to 65°F using a heating coil while keeping the scions cool (50°F to 55°F). During this period, it's essential to monitor for fungal disease, periodically applying fungicides if required. Provide lots of indirect lighting.

- Once the scion begins to grow, after approximately three weeks you'll want to provide ventilation by removing the plastic covering for a few hours every morning to help acclimate the grafts to lower humidity. At this time, remove about one-third of the shoot portion of the rootstock, particularly the succulent shoots. Discard any dying or dead grafts. If the graft union has already callused, you may be able to slightly loosen the rubber tie to allow the graft to increase in width.
- After another four weeks, if the scions are growing well, you can trim back most of the rootstock. Retain one or two shoots of the rootstock to provide additional photosynthesis to support the roots and the young scion.
- By spring, the plastic covering can be removed from the strongly growing grafts and the containers placed in the shade or in a well-ventilated greenhouse for further acclimation. After a few weeks, move them outside, ideally into afternoon shade. After a few more weeks, they can be exposed to full sun. At this time, remove the less vigorous rootstock shoots.
- The following spring remove the remaining rootstock shoots, even for the smaller, slower-growing varieties. The new growth on upright cultivars usually needs to be staked upright.

Propagation of witch's brooms by grafting has been a great source of dwarf and miniature conifers. The degree of increased branching and dwarfism varies considerably, with some growing less than 1 inch per year and others growing 2 to 6 inches per year. The growth form can also vary, taking the form of a ball, mound, narrow upright column, prostrate spreader, or weeper. Grafting of scions from a witch's broom onto a normal rootstock of the same species or related species of the same genera usually develops the special attributes of the broom.

Rising above the iconic moon bridge in Seattle's Kubota Garden, a slender Serbian spruce reaches for the sky.

In the next chapter, we'll review how to keep your conifer thriving in a healthy garden ecosystem.

Sustainably Caring for Conifers

Chapter 8

Soil

Healthy soil is the foundation of a sustainable landscape and a vibrant ecosystem. Healthy soil decomposes organic matter, sequesters carbon, captures nutrients, makes those nutrients useful to plants, maintains soil structure, and keeps soil organisms in proper balance. It's essential to understand your garden's soil type, structure, organic matter content, and overall soil health—including its depth, degree of compaction, drainage, texture, moisture content, and pH (acidity or alkalinity). Ideally, your comprehensive site evaluation (discussed in Chapter 3) yielded this information.

A healthy loam soil usually contains 3% to 7% organic matter and approximately equal proportions of silt, sand, and clay. However, clay soil, with its beneficial nutrient- and water-holding properties, is fine for nearly all conifers. I have successfully grown more than 130 different species of conifers, including many drought-loving pines, in the predominantly clay soil of my garden.

To mitigate drainage problems from clay soil,

1. Dig the soil to a depth of at least 1 foot (at a time when it is neither too wet nor excessively dry).
2. Add compost. If the soil is light-colored or rich in clay, add a 3- to 4-inch layer of fine compost, and dig it into the existing soil to create more pore space.
3. Add 3 inches of coarse organic mulch on top of the soil.

If, on the other hand, your soil contains too much sand and therefore drains too fast (which allows roots to dry out), you can mix fine compost into the sandy soil and top dress with mulch to enhance moisture retention.

If water pools too long in your landscape after heavy rain, you can facilitate stormwater infiltration and dispersion by replacing impermeable surfaces—such as the driveway, walkways, or patios—with permeable surfaces. You can also redirect the water from your downspouts to one or more rain gardens.

Organic Matter

Adequate organic matter is essential for soil health and growing healthy plants. Organic matter serves as a reservoir of nutrients and water, reduces compaction, increases water infiltration, and supports beneficial soil organisms. Sustainably manage your garden and kitchen waste by composting it into a (free) soil amendment. Arborist's wood chips are an excellent mulch for conifers and other woody plants. They diffuse rainfall; reduce soil erosion, compaction, and surface crusting; moderate soil temperature; prevent weed germination; neutralize urban pollutants; and improve soil structure, aeration, and porosity. Truckloads of wood chips are often available for free from local tree service companies. If you cannot use a full truckload, share the wood chips with your neighbors.

Mycorrhizae

Many conifers have primarily fibrous roots that grow in the upper layer of the soil; most lack a deep taproot. Having a shallow root system to supply the tree with water and nutrients makes it easier for conifers to grow on poor, thin, or rocky soils. They get help in acquiring water and nutrients by forming a mutualistic association with specialized fungi, called *mycorrhizae* (from the Greek words *mycor,* for "fungus," and *rhiza,* for "root"). Mycorrhizae have tubular filaments, or *hyphae,* which interact with the conifer's fine roots. In return, the conifer's fine roots exude carbohydrates and other carbon-containing compounds, feeding the mycorrhizae. There are two different kinds of mycorrhizal associations: *ectomycorrhizae* and *endomycorrhizae.*

1. *Ectomycorrhizal associations* are seen in most members of the pine family (Pinaceae). They feature *mycelia*—highly branched, threadlike strands of hyphae growing together—which surround and form a dense sheath around the plant's feeder roots. These hyphae grow between the conifer's outer epidermal cells and cortical cells, forming a network called the *Hartig net.* The fungi, with its hyphae sheath and Hartig net, produce the plant hormone *auxin,* which makes roots grow short and thick with minimal branching, taking on a distinct club-like appearance. Most of the fungal partners are *basidiomycetes,* which produce aboveground fruiting bodies (mushrooms) that have wind-disseminated spores.
2. Most other Pinaceae have *endomycorrhizal associations,* in which the fungal mycelia is confined to and actually penetrates the cortical root cells (the cortex), producing vesicles and branched clumps called *arbuscular mycorrhiza.* Thus, endomycorrhizae are called *vesicular arbuscular mycorrhizae.*

Mycorrhizae not only improve the absorption and transportation of nutrients and water, making conifers more tolerant of extended dry periods, but they also sequester carbon, improve soil structure, and help hold soil together through hyphal networking and *glomalin* (biological glue) production. They may also enhance resistance to biotic stresses—such as herbivory, insect feeding, and fungal pathogens—and abiotic stresses, such as mechanical wounding.

Mycorrhizae spores are present in most native soils, although overwhelmingly sandy soils generally lack mycorrhizal associations. Altered or degraded urban soils and soils in which the topsoil layer has been removed might lack mycorrhizae, and a store-bought conifer might not have mycorrhizal root associations. Sometimes, young conifers are inoculated with mycorrhizae during the seedling and nursery stages to promote their survival. If your conifer has not been inoculated, and you think you have mycorrhizae-deficient soil, you can inoculate your soil with native soil from a nearby area with healthy trees and shrubs. Alternatively, you can inoculate your soil with the appropriate commercial mycorrhizae when you plant it, following the commercial supplier's instructions.

Cultural Practices

Fertilizing

Conifers use nutrients more efficiently than most hardwood deciduous trees because they retain most of their leaves for several years, annually replacing less than 20 percent of their foliage. Therefore, conifers generally require fewer nutrients than hardwood deciduous trees. In fact, fertilizing conifers can be detrimental because excess nutrients can promote excess soft new growth susceptible to insect attack. Consequently, most conifers growing in good soil do better without added fertilization. Interestingly, in the Northwest, conifers get a significant proportion of their nutrients from their own needles: during fall and winter, seasonal rains leach high levels of nutrients out of the leaf litter and top layer of soil down to the root zone, where they can be taken up by roots. In contrast, most Northwest hardwoods get their nutrients primarily in late spring and during the dry summer months, when nutrient availability is relatively low. This is one reason why conifers adapted to wet winters and dry summers dominate the Pacific Northwest.

In gardens, healthy soil can usually provide adequate macronutrients (nitrogen, phosphorus, potassium, sulfur, and magnesium) and micronutrients (iron, boron, manganese, molybdenum, copper, zinc, and chlorine) to conifers. (However, in special situations involving volatilization [conversion of organic nitrogen to ammonia gas] or denitrification [conversion of nitrates to gaseous forms of nitrogen], soil nitrogen can become deficient.) I generally do not add fertilizer to the garden. I find that mulching with 3 inches of coarse mulch once a year usually provides enough nutrients. For soil depleted in organic matter, it may be necessary to incorporate about 3 inches of fine compost into the soil as well. Soil organisms capture the nutrients in these soil amendments and convert them to a usable form. Nutrient-poor sandy soils may require the occasional addition of slow-release fertilizer.

The optimal pH for most conifers is between 5.5 and 6.5, which is slightly acidic. Some conifers, such as junipers, can tolerate basic soils (above pH 7.0), and others, such as arborvitae (*Thuja occidentalis*), false-cypress (*Chamaecyparis* species), and white spruce (*Picea glauca*), can tolerate a pH near neutral (pH 7.0). For most conifers, having insufficiently acidic soil can cause manganese and iron deficiencies. Too-acidic soil can cause phosphorus deficiency. Sandy soils, especially those with low organic matter, don't hold nutrients well and can be deficient in one or more soil nutrients.

To head off nutrient problems, first assess the soil's pH and nutrient content using a certified soil test, which you can find through your local Extension office or Master Gardener clinic. It is important to collect multiple soil samples from the area in which your conifer will grow. To identify a specific nutrient deficiency, you may also need a foliar test, which assesses a conifer's leaves.

You can fix most nutrient deficiencies by adding recommended soil amendments in spring, when the conifer is actively growing. However, if you need to correct a potassium deficiency by adding potassium chloride, do it in the fall, so the fall rains can wash away excess chloride. Add only deficient nutrients in amounts determined by the testing agency. Never add excess nutrients to the soil. They do not benefit the plant, and they harm the environment. Nitrogen and phosphorus, in particular, readily leach from the soil and pollute nearby waterways, stimulating algae blooms and killing fish in a process called *eutrophication*. Too much soil nitrogen also stimulates aboveground growth rather than root growth, making conifers especially susceptible to drought. The root system can grow too little relative to the leaf area to provide adequate water to the canopy during dry periods. High nutrient levels can further reduce the conifer's drought tolerance by inhibiting mycorrhizal associations.

During photosynthesis, the pores on conifer leaves take up carbon dioxide and release oxygen and a small amount of water vapor; the process is called *transpiration*. Transpiration creates a negative pressure that pulls water up from the soil through the tree's *water column*, a transport system that runs from the roots through trunk, branches, and twigs all the way to the leaves. If the soil is too dry, water evaporating from the leaf pores creates tension in the water column. After many hot, dry days, the water column can break—a phenomenon called *cavitation*—creating an air bubble and disconnecting the roots from the leaves. After prolonged lack of water, the tree's water transport system stops working altogether, and the top, branches, or even the entire tree can die.

Minimizing Water Stress

Though conifers may be less vulnerable than hardwoods to drought, you'll still want to minimize water stresses that make conifers more susceptible to pests and diseases by doing the following:

- Do not plant trees on excessively fast-draining sites or in very shallow soils.
- Add enough organic matter to your soil to ensure adequate water-holding capacity and facilitate mycorrhizal associations.
- Plant in native soil with existing mycorrhizal associations.
- Provide plenty of space, so the conifer won't compete with other plants for water and nutrients.
- Remove all grass from between the trunk and the drip line of the tree, as grass can divert water from the tree's roots.
- Mulch with 3 inches of coarse wood chips around the tree, over the roots, and to the dripline to retain moisture, keep the roots cool, and avoid soil compaction. Keep mulch a few inches away from the trunk, however, as it can rot the bark.

- Plant a diversity of plant species, including a diversity of flowers, beyond the conifer's dripline to attract beneficial insects to reduce problem insects and disease.
- Provide afternoon shade for conifers, especially if you live in a warm region. Many conifer cultivars with variegated or yellow foliage need protection from hot afternoon sun. Remember, very dry summers are common on the West Coast.
- Provide infrequent, but deep, watering during dry periods.

Weeding

To reduce weeds, plant dense groundcovers and apply a coarse mulch. Get to know the common weeds and their life cycles so you can remove them before they go to seed and spread.

Pruning

Proper pruning helps maintain your conifer's good health and desirable form. Before you start pruning, become familiar with your conifer's foliage, cones, movement, balance, and proportion throughout the seasons. Learn its horticultural needs and how it responds to pruning cuts.

Conifer expert Sam Pratt pruning out a dead branch on gracefully weeping dawn redwood 'Miss Grace'.

Conifers have two pruning groups, based on branch arrangement:

1. Those with whorled branches, which form a circle around the growing leader—such as pines, spruces, true firs, true cedars, larches, and Douglas-firs—have one flush of growth each spring from buds formed during the previous year's growth. New growth comes from the visible buds on the current season's new growth; they don't have latent buds on old wood. Typically, the one flush of new growth develops in a circular pattern at the tips of the branches and the top of the central leader.
2. Conifers with a random branching habit—including arborvitae, false-cypress, cypress, yew, juniper, and hemlock—have multiple growth spurts throughout the growing season from many latent buds along their stems.

By selecting an intermediate-size or dwarf conifer for an urban garden, you will minimize the need to prune for size. You can, however, enhance the health and reveal the natural beauty and character of your specimen through aesthetic pruning.

Conifer expert Sam Pratt demonstrates how you can enhance the health and beauty of your conifer with a bit of aesthetic pruning, helping to maintain its shape and balance as it grows. Pictured is a Japanese larch *Larix kaempferi* 'Paper Lanterns'.

Remove dead, diseased, or damaged branches as soon as they appear. Make a clean pruning cut flush with the outside of the swelling where the branch attaches to a larger branch, called *the branch collar.* Cutting at the outside of the collar allows the conifer to grow wood over the cut area and seal it. Cutting into the xylem invites fungal infections and destroys regenerative cells located in the collar. Cut at a slant to enhance runoff. Remove any shoots of a cultivar that aren't true to that cultivar—for example, remove an all-green shoot on a variegated cultivar. Also remove shoots growing from the rootstock of a grafted conifer. Otherwise, the reverting shoot or rootstock will eventually take over the plant, and you will lose the cultivar's unique characteristics.

To enhance air circulation and light coming through the conifer, thin some of the branches in winter or early spring, when the conifer is dormant and the weather is not wet or exceedingly cold. On large conifers, you may remove some lower branches, and thin a small portion of the upper branches. Even if you've selected a conifer specifically to fit an allotted space, you might want to curtail its growth, encourage fullness, or improve its shape. On whorl-branched conifers, prune some growing tips in mid- to late spring, when the plants are actively growing. To control growth in pines, snap off a portion of the developing bundles of needles, called *candles,* with your fingers. The more of the candle you remove, the slower new growth will be. For other species with whorled branching, prune just above visible buds or branch internodes where green needles are present. Prune in late winter or early spring before new growth starts. You can also prune in mid-summer when the conifer is semi-dormant. Where the pitch moth is active, such as on the West Coast and in parts of the Rocky Mountains, minimize pruning from April to October, when the adult pitch moth is laying eggs, as new pruning wounds attract moths.

On a randomly branched conifer, remove branches that cross other branches or ruin the form, and prune almost anywhere along the stem but not into bare wood. Prune in early spring so new growth hides the pruning cuts. To maintain the conifer's size, prune it lightly in summer, when conifer growth is minimal.

To reduce the width of conifers that have random branching, such as some false-cypress or arborvitae, when the tree is young, reduce the number of leading shoots every few years in midsummer. It's easy to control the size of some of the randomly branched conifers, such as yews and hemlocks, because they have abundant buds on old and new wood. These make good hedges because they readily grow at the pruned sites and form dense canopies. Prune conifer hedges in spring, just before new growth begins, so new growth covers the pruning cuts. For junipers, arborvitae, and false-cypress, whose leaves are intolerant of shade, don't cut beyond the leafy area because they will not leaf out from the old branches. For some variegated conifers that tend to lose their variegation with age, such as *Tsuga canadensis* 'Gentsch White', prune off the branch tips every few years to stimulate variegated new growth.

To increase the leaf density of spreading or prostrate conifers, such as some junipers and yews, prune the shoots in early spring before new growth begins. Topiary—pruning and training conifers into unusual or unnatural shapes—has its aficionados. I personally much prefer the natural shapes of conifers, and there are myriad shapes, sizes, and colors to choose from. A final note: Be sure your pruning tools are clean and sharp, and disinfect your tools with a mild, yet effective disinfectant, such as Lysol. Also, be sure to disinfect your pruners after removing diseased portions of your conifer.

Some conifers help protect themselves from pests and grazing mammals by oozing large amounts of sticky pitch. Pictured is a lodgepole pine 'Chief Joseph'.

Conifers' Protective Strategies

To protect themselves from insect pests, conifers have developed physical and chemical defense mechanisms. Many conifers produce resins and other complex volatile organic compounds, such as *sesquiterpenes* and *monoterpenes,* that repel or kill stem-borers and foliage feeders or that attract predators or parasites of those pests.

Humans have used resin for thousands of years. Amber, for example, is fossilized plant resin; artisans still use amber from now-extinct pine trees to make jewelry and art. Shipbuilders have used resin to waterproof their boats and ropes, and manufacturers today distill pine resin to yield turpentine and rosin, which have many practical uses.

Many conifers have resinous leaves and *resin canals* or *resin blisters* on young stems, older branches, and tree trunks. Resin canals are long-lasting, highly organized systems of interconnected tubes lined with specialized epithelial cells that produce resin. For example, when ponderosa pines are water-stressed, epithelial cells lining the resin ducts become flaccid, and the oleoresin pressure can drop to zero. Bark beetles target these trees because the trees aren't producing their protective resin. Those conifers with few resin ducts, such as lodgepole pines, or no preexisting resin ducts, have a *passive resin system.* A passive resin system produces *traumatic resin* in response to wounding, which floods invading insects with viscous and toxic terpenoids. When the resin is exposed to air, it hardens, sealing the wound against insects and fungal pathogens or trapping and poisoning them. Several fir species, such as *Abies pindrow* and *Pinus strobus,* have seed cones that excrete resin out of their tops to deter herbivores. True firs, cedars, hemlock, and golden-larch all have simple and short-lived resin blisters. These resin-filled blisters are ubiquitous on the trunk of balsam fir. Juniper, arborvitae, and cypress leaves have resin glands on their undersides. Bald cypress has resins in its seed scales. Resinous and aromatic plants, including eucalyptus, rosemary, and many conifers, are highly flammable. If you live in an area at high risk for wildfires, such as near woodlands in the West, avoid using high-resin conifers, and choose fire-resistant plants.

Integrated Pest Management

Integrated pest management (IPM) is a scientific approach to the prevention, monitoring, and management of pests in the garden while minimizing environmental and health risks. How susceptible your conifers are to disease and pests depends on how well-suited they are to their garden conditions; pathogens and insects seldom seriously affect healthy plants.

Integrated pest management involves the following key tasks:

1. Determine the cause of any damage, and properly identify the host and the responsible pest.
2. Regularly monitor for pest activity.

3. Set thresholds for how much pest damage you are willing to tolerate.
4. Establish a plan of action for dealing with a pest problem.
5. When necessary, take effective, timely action, using the least-toxic method to control your pest.

Regular monitoring is an essential part of integrated pest management. Catch damage as soon as you can because the pest that caused it may be on the tree only briefly. Even if the pest itself is absent, look for insect signs:

- silk larval cases
- trails of silk
- frass (insect excrement)
- scale and aphid exoskeletons that were shed during molting
- honeydew (a sugar-rich liquid secreted by pests such as aphids and soft-bodied scales)
- black sooty mold growing on the honeydew
- insect parts
- scale eggshells
- larval skins
- pupal cases and cocoons
- damage caused by the tree's reaction to the insect wounds

Tent caterpillars eating on a conifer in the photographer's garden in West Seattle.

You will not be able to identify all the insects that attack conifers, but by identifying the damage pattern, you can narrow your suspects. For example, beetles (including weevil) and moth larvae are terminal branch feeders.

Sequoia pitch moths (*Synanthedon sequoiae*) and Douglas-fir pitch moths (*Synanthedon novaroensis*) drill under the outer bark of susceptible pines, spruces, and Douglas-firs in the western states. The larvae produce large, ugly pitch masses on tree trunks and branches, particularly near limb junctions and pruning wounds or other injury sites. Look under the pitch for dirty-white, greyish, or pink larvae or dark brown pupae. Monitor for reddish-brown to pink masses of resinous ooze that harden to dark gray. Carefully scrape away the resinous mass, and then tease out and kill the larvae. Unsightly branches with pitch masses can be pruned off when the tree is dormant, in late fall or early winter, cutting just outside the branch collar to promote sealing off the wound before egg-laying adults emerge in spring. To keep moths from reproducing, you can put sticky traps baited with the sequoia pitch moth pheromone Alpha Scent in susceptible conifers; that way, you will trap the male moths. Physically removing the resin masses when the larvae are still present and destroying the larvae can effectively reduce reinfestations. For cultural control, infrequently but thoroughly irrigate during dry periods (but avoid overwatering), because water-stressed trees are more susceptible to the moth. Also, protect the trees from injury.

Douglas-fir tussock moth (*Orgyia pseudotsugata*) is a western moth that defoliates Douglas-fir (*Pseudotsuga menziesii*), spruce (*Picea* spp.), and true fir (*Abies* species) growing east of the Cordilleran Mountain Range. True firs are their favored hosts. The hairy larvae hatch from the overwintered eggs just after bud break. They start by feeding on new needles then move to older ones, so the damage pattern is from the top of the tree down. At first, you will see messy, discolored needles and later heavy defoliation, followed by bark beetle attacks. Natural enemies usually control tussock moths after a season or two. *Note:* Limit *your own* contact with the caterpillars, as they can cause allergic reactions, including itching and rashes.

Sawflies (in the *Neodiprion* genus, the *Diprion* genus, and a few other genera) are primitive wasps. Females cut slits in needles or scales to lay their eggs. The larvae first feed in unison, and later individually, on pine, spruce, or larch trees. Some sawfly species are also leafminers, such as the lodgepole pine leafminer. Leafminers rarely kill trees, but they defoliate their hosts and make them look unsightly. Sawflies generally feed on old needles, discoloring foliage and leaving needle stubs. Sawflies feed on most conifer species, though each sawfly species has its own predilections. Outbreaks are cyclical, but management of sawflies is rarely necessary because freezing weather or native parasites, along with naturally occurring fungi and viruses, usually control them.

Adelgids, aphids, and scale insects are sap-sucking defoliators. These organisms use straw-like mouthparts to pierce the phloem and feed on the sugar-laden sap. This destroys the needles, causing needle distortion, necrosis, and needle drop, and leads to dead spots in the foliage and thinning crown.

Hemlock woolly adelgids (*Adelges tsugae*) are aphid-like insects. When they were introduced to the East Coast from Asia, these insects had no natural controls and rapidly spread. Hemlock woolly adelgids feed on the sap of hemlocks, including Canadian hemlock (*Tsuga canadensis*), western hemlock (*T. heterophylla*), and the Carolina hemlock (*Tsuga caroliniana*) in the Appalachians. The saliva of these pests is toxic to meristematic tissues (those that have the ability to actively divide throughout their life) in the terminal buds and other stem-expanding tissues of trees. Interior needles turn yellow and drop prematurely, killing limbs and even, sometimes, the tree. Look for white woolly tufts of waxy threads on the undersides of stems. Fortunately, woolly adelgids cause nominal damage to the western hemlock (*Tsuga heterophylla*).

Cooley spruce gall adelgid (*Adelges cooleyi*) is native to the Rocky Mountain region but introduced all over the world. These adelgids feed on new growth around the base of needles in an expanding bud, making it swell into a green gall. The galls appear at the tip of the shoot. Douglas-fir is a secondary host for *A. cooleyi*. On Douglas-fir, Cooley spruce gall adelgids attack the foliage, producing distorted, discolored needles. On spruces, prune off and destroy the galls in early summer before they open and the adults emerge. You can also apply horticultural oil before new growth starts—when the temperature is above freezing but the conifer is still dormant. Note that the oil treatment removes the glaucous bloom of blue spruce needles, but the discoloration is temporary.

Balsam woolly adelgids (*Adelges picea*) are introduced European pests whose lifecycle is spent entirely on true firs. Today, these female adelgids, which emerge from unfertilized eggs, occur in the northwestern and northeastern United States and across southern Canada. In Europe, part of their lifecycle is on spruce, where males are produced. In North America, overwintering balsam woolly adelgids look like white tufts on branches and leaders, but they appear all over the trunk in advanced infestations. These female adelgids develop explosive populations and cause swellings of new buds and branchlets, killing firs from the top down. They have devastated vast swaths of subalpine firs along enormous areas of the Cascade Range and the Sierra Nevada Mountains.

Spruce aphids (*Elatobium abietinum*), another introduced European pest, feed during winter months on old needles of native and ornamental spruces. Look for small green aphids on the undersides of older yellow needles as spring approaches, especially where winters are mild. Aphids thrive on foliage high in nitrogen—a good reason to avoid nitrogen fertilizers on spruces. You can wash the aphids off ornamental spruce trees with a strong water stream from a hose.

Pine needle scale (*Chionaspis pinifoliae*) is an armored scale insect that attacks ponderosa pine (*Pinus ponderosa*), lodgepole pine (*Pinus contorta* var. *latifolia*), and other one- to three-needle hard pines, slowing growth and weakening trees. Unlike soft-bodied scales, *C. pinifoliae* does not produce honeydew. Instead of feeding on sap, these scales feed on mesophyll cells in the needles, which generate plant sugars through photosynthesis. So, you won't find black sooty mold associated with armored scales. Look, instead, for these scale insects feeding on needles. You can detect the tiny crawlers by shaking a branch over a piece of paper. Control pine needle scale by applying horticultural oil during the dormant season or by spot spraying with insecticide in spring, when the scale is in its vulnerable crawler stage.

The larvae of white pine weevils (*Pissodes strobi*) tunnel under the bark of branch tips in late spring, girdling and killing branch tips by summer. White pine weevils damage ornamental pines and spruces, making them unattractive and bushier, with multiple leaders. Both white pine (*Pinus strobus*) and spruce (*Picea* spp.) are the weevils' main hosts in eastern North America, with Norway spruce being especially vulnerable. In the West, Sitka spruce, Engelmann spruce, Colorado spruce, Norway spruce, and Serbian spruce are the preferred hosts. The white pine weevil also attacks jack pine, red pine, and Scots pine in the prairie regions of Alberta, Saskatchewan, and Manitoba.

Adult weevils become active in spring when daytime temperatures consistently exceed 50 degrees Fahrenheit. They feed in sunny habitats on the bark of healthy, vigorously growing trees 6 to 25 feet tall. Oozing pitch on the top leader is a symptom of weevil feeding and egg-laying. A discolored drooping or wilting terminal leader is another, along with larval mining under the bark. To control white pine weevil, monitor susceptible trees weekly in spring. As soon as you notice damage, prune out larvae-infested leaders just above the topmost unaffected whorl of branches. Peel back soft bark to look for weevil larvae or tunneling. Choose a vigorous side shoot near the removed leader and tie it upright; it will become the new leader. Prune off the other shoots in the whorl of branches just below the cut terminal. Destroy the pruned shoots by burning or seal them in a trash bag and dispose of them in municipal waste.

An alternative approach is to wrap the base of the trunk and the top of the leader with sticky tape to trap weevils crawling up the trunk. You can also place Tedder traps (made by Great Lakes IPM) near previously infested trees; the traps contain ethyl alcohol and turpentine, which attract weevils, allowing you to monitor for them and help control populations. If mechanical controls are insufficient, drill a ⅜-inch hole in the trunk, and insert an acephate systemic insecticide tree implant. Alternatively, and as a last resort, a licensed pesticide applicator can spray insecticides in early spring to kill overwintered adults before they lay eggs.

Most twig beetles (*Pityophthorus* spp. and other tiny bark beetle genera) produce larvae that tunnel under the bark of pine, spruce, true firs, and Douglas-fir branches, wilting the terminal growth. Look for tan or reddish sawdust as a sign of infestation. Newly transplanted pines and pines that are stressed by drought or by root or shoot damage are particularly susceptible. There is no treatment once the tree is infested, so minimize stress by providing optimal cultural practices.

Brown leaves or needles can indicate arachnids such as spider mites as well as activity by *homopterans,* herbivorous insects with piercing-sucking mouthparts and a sap-sucking mode of life; this group of pests includes aphids, scales, treehoppers, cicadas, adelgids, and many others. Tap a branch over a piece of white paper or look at a branch with a magnifying glass to verify their presence. You may be able to remove many of the sapsuckers with a strong jet of water.

If you can't identify the pest, take a damaged section of your conifer, along with the pest, if available, to your local Extension office, Master Gardener clinic, or local entomologist. You must know the pest before you treat.

Next, learn about the pest's major characteristics, lifecycles, and management options to determine whether you should tolerate the pest. If you decide to act, use cultural, physical, or biological controls before using chemical control. Always select the most-practical yet least-hazardous approach.

Cultural controls involve choosing pest- and disease-resistant plants appropriate for the specific conditions of climate and microclimate of your site as well as properly watering and correctly fertilizing to keep plants healthy and robust. If the wrong plant is in the wrong place, it should be transplanted to a new site or removed.

Physical controls involve directly removing or killing pests—for example, by hand or with a jet of water—or physically blocking the pest from the plant by using traps or other physical barriers.

Biological controls involve attracting a pest's natural predators, parasitoids, or pathogens (disease-causing agents) to your garden by planting a diversity of nectar-rich flowering plants offering nectar and pollen throughout the growing season. If necessary, biological controls can be purchased commercially and released into the landscape to augment the pest's natural enemies as a preventive measure to provide long-term control or to manage and treat a pest outbreak. Knowing which species to purchase requires the correct identification of the source of the damage. To select the most appropriate control agent, learn about the life cycles of the pest and its natural enemies as well as the optimal time to release the control agent. Biological control augmentation is most appropriate when the damage from low levels of pests can be tolerated. If these approaches don't work, try an alternative tactic or a combination of tactics. Then, if necessary, escalate to using the least-toxic chemical control and application method at the most-effective time. Keep a record of control measures, timing, and effectiveness.

Dealing with Deer

Deer prefer edge spaces, at the interface of openings and areas of cover—for example, where green spaces abut residential development. In these urban and suburban areas, the food supply for the deer is stable and abundant, hunting is rarely allowed, and predators such as wolves, cougars, and coyotes may not be hunting deer. Because of this, deer populations have surged in America's urban and suburban gardens. Deer significantly damage garden vegetation when they browse or eat the growing tips of shrubs and trees. Bucks seriously damage conifer bark and cambium when they

Mule deer and piñón pines near Great Basin National Park in Nevada.

scrape the velvet covering from their antlers against the flexible stems of young saplings; they can even kill saplings when the cambium is completely removed around the trunk. Serious damage is more likely if the main trunk is exposed, which I found out the hard way when a black-tailed buck selected my columnar pine saplings for scraping his antlers.

When food sources are scarce, hungry deer will eat almost anything in the garden. The end of winter or a long dry spell is usually the worst for deer damage. The best solution is probably to plant deer-resistant plants. In general, deer do not like plants with aromatic, leathery, spiny, or hairy leaves, but it really depends on the species of deer and what kind of vegetation is otherwise accessible to them. Deer do not usually browse on most conifers, although they are known to like yews (*Taxus* species) and eastern arborvitae (*Thuja occidentalis*). The white-tailed deer (*Odocoileus virginianus*), Columbian black-tailed deer (*Odocoileus hemionus* subsp. *columbianus*), and Rocky Mountain mule deer (*Odocoileus hemionus* subsp. *hemionus*) each have different preferences. In the western United States, all three browse Douglas-fir (*Pseudotsuga menziesii*). Both black-tailed and white-tailed deer browse western-red-cedar (*Thuja plicata*). Mule deer common to the region east of the Cascade Mountain Range, including the western Great Plains, Rocky Mountains, and the US Southwest, browse ponderosa pine. Unfortunately, no conifer is 100 percent deer-proof. For a list of deer-resistant plants for the eastern United States (including conifers), see https://njaes.rutgers.edu/deer-resistant-plants/. For the western US, see https://emswcd.org/wp-content/uploads/2013/10/OSU-Deer-Resistant-Plants.pdf.

Solutions for deer in your garden include the following:

- **Erecting a barrier,** such as an eight-foot deer fence. In many urban areas, however, fences can be no taller than six feet, which deer can jump. They are much more likely to do so, however, if they can see through the fence, so a solid six-foot fence would be a reasonable deer deterrent, although that may be too obtrusive for some homeowners. I've found that a six-foot see-through fence can deter deer if a dense mix of vegetation blocks the deer's view on the other side. Alternatively, you can add a second fence about four feet from the first or use an electric fence. To protect an especially vulnerable or valuable conifer, surround it with a five-foot fence far enough away so the deer cannot reach it with their long tongues. Be sure to tightly secure the fence to the ground with pegs, so the deer cannot move it or crawl under it.
- **Keeping a large dog** in the yard to chase away the invaders.
- **Using a fragrance or a deer repellant.** This is usually only a temporary fix, and you need to reapply it regularly, particularly after a heavy rain.

Looking out from the author's garden, we see a hungry, black-tailed doe who would undoubtedly love to trade places, peering in through chain-link deer fencing.

Invasive Plants

Non-native, Himalayan blackberries left untended can completely overwhelm a landscape.

Invasive plant species abound in North American urban ecosystems, as human activity spreads them widely. These aggressive plants compete for resources and space, often overtaking native and nonnative conifers, other ornamental trees and shrubs, and beneficial understory plants. This results in plant communities homogenizing. Urban biodiversity decreases, damaging the urban tree canopy and the urban forest. For example, the following three plants are some of the worst invasive species west of the Cascade Mountain Range in the Pacific Northwest:

- Invasive ivies, especially English ivy (*Hedera helix*) and Atlantic ivy (*Hedera hibernica*) quickly climb trees, blocking sunlight and preventing photosynthesis. They can also add thousands of pounds of weight to a tree, which can snap them or cause windthrow.
- Himalayan blackberry and invasive knotweeds outcompete the native understory for sunlight, water, and soil nutrients, weakening surrounding plants and preventing them from reproducing, either from seed or by suckering.

To protect your garden from invasive plants, learn to identify them and remove them. Replace them with site-adapted native or nonnative plants that benefit you and the urban wildlife. Volunteer to help remove invasive species from your neighborhood. You can find groups working on this problem through your County Noxious Weed Board or through your County Extension office. This is an ongoing (some gardeners might say never-ending) process, so be vigilant about watching for and immediately removing invasive species.

Conifer Diseases

Most conifer diseases are caused by true fungi, fungus-like organisms (phylum: *Oomycota*), or dwarf mistletoes, which are parasitic. The interactions between hosts and pathogens play an important role in natural forest ecosystems and their resilience in a changing environment. Strong, healthy trees are not significantly affected by pathogenic fungi, but trees weakened by abiotic factors such as wounding, unusual weather, or climate change are more susceptible to damage. Properly caring for your conifers can limit diseases. Although many species of fungi and other pathogens can cause disease, we will focus on the most common found on conifers found in urban landscapes and nurseries.

Foliage Diseases

Rhizosphaera needle cast disease is a fungal disease of spruces and other conifers in the Midwest and Northeast. It causes early needle loss of the innermost needles of the lower branches, particularly on Colorado spruce (*Picea pungens*). However, other spruces, such as Engelmann (*Picea engelmanii*), Austrian (*Picea nigra*), Serbian (*Picea omorika*), and Sitka spruce (*Picea sitchensis*) as well as Douglas-fir (*Pseudotsuga menziesii*) and western hemlock (*Tsuga heterophylla*) can succumb, too, if the trees are stressed.

Newly emerging needles in spring are infected by airborne spores, but symptoms are most apparent in older needles. If your spruce has discolored yellow or tan older needles toward the bottom of the tree, there is a good chance it is infected with the fungus *Rhizosphaera kalkhoffii*, or a closely related species. This pathogen overwinters in needles on the tree and on the ground. As the disease progresses, the infection moves up the tree to infect new needles, particularly if there have been extended periods of moisture.

To determine whether your tree has Rhizosphaera needle cast, examine the undersides of the infected needles with a 10× hand lens. If you see rows of black dots along the length of the needles, your tree is infected. These dots of small fruiting bodies contain spores that spread the disease. To minimize the chance of infection take the following steps:

- Remove and destroy the oldest needles as soon as they change color.
- Prune lower branches with localized infections during dry weather to enhance airflow through the tree and quicker drying of needles. Best to disinfect your pruners between each cut.
- Remove and destroy infected needles on the ground.
- Irrigate trees at the roots. Don't use overhead irrigation, which wets the needles.
- Treat the tree with a copper-containing fungicide or chlorothalonil, alternating these treatments every three to four weeks. Before using any pesticide, always read the label directions and follow them carefully.

Diplodia tip blight is a common fungal disease of two- and three-needle pines, particularly in New England, the Midwest, and the West. It affects Austrian (*Pinus nigra*), Scots (*P. sylvestris*), and red (*P. resinosa*), ponderosa (*P. ponderosa*) and mugo (*P. mugo*) pines. Interestingly, the nonnative Austrian pine is more likely to become severely infected than native pines. Diplodia tip blight also can infect many other conifers, including eastern white pine (*Pinus strobus*), Douglas-fir (*Pseudotsuga menziesii*), Norway spruce (*Picea abies*), Colorado spruce (*Picea pungens*), American larch (*Larix laricina*), noble fir (*Abies procera*), true cedars (*Cedrus* spp.), arborvitae (*Thuja* spp.), and junipers (*Juniperus* spp.), particularly if they are weakened by abiotic (nonliving, such as temperature extremes or lack of water) or other biotic factors.

Diplodia tip blight is caused by the spores of the fungus *Sphaeropsis sapinea* (formerly *Diplodia pinea*), which spread mostly by wind, sometimes by rain, and primarily attack the current season's buds and shoot tips. Symptoms include brown, stunted, and/or wilted new shoots with short, brown needles. Strong young trees can confine infections to new shoots, but seedlings and trees over 30 years old tend to fall prey to this blight. It can also attack older stems that are wounded by pruning, insect feeding, and snow or hail. Weaker ornamental trees, including those growing in unfavorable cultural conditions, can also be seriously affected.

Symptoms usually appear first on the lower portion of the tree and progress upward, killing all needles on the current season's shoots. A general dieback occurs in the canopy near the branch tips, and you can often see resin droplets and discolored underlying wood on the dying or dead shoots. If you look carefully in the summer and fall, you will see black-pepper-like fruiting bodies containing spores called *pycnidia* at the base of the needles under the *fascicle sheath*, the structure that clusters needles, or on the scales of second-year cones. To treat this blight take the following steps:

- Prune and remove all blighted shoots and stems.
- Avoid fertilizing with excessive nitrogen because it worsens Diplodia disease.
- Apply a fungicide effective against Diplodia at least three times: at budbreak, at half candle, and again at full candle. Do this yearly until the infection is contained or minimal.

Cytospora canker of spruce is caused by the fungi *Cytospora abietis* (syn. *Valsa abietis*) and *Cytospora kunzei* (syn. *Leucostoma kunzei*). They affect many species of spruce grown outside their native range, especially Colorado spruce (*P. pungens*) and Norway spruce (*P. abies*), as well as Engelmann (*P. engelmanii*), Norway (*P. abies*), and red (*P. rubens*), white (*P. glauca*), and black spruce (*P. mariana*). Many other conifers can also be attacked by Cytospora canker.

The fungi primarily attack trees over 15 years old and trees under stress. First symptoms of Cytospora canker include dying lower branches and browning needles. As the disease progresses, whitish resin patches often appear on the bark of branches with cankers. As the cankers grow larger and girdle the branch, they disrupt the food-conducting (phloem) and mineral-and-water-conducting tissue (xylem), causing the branch above the diseased area to wilt and die. If you carefully remove a thin layer of bark, you'll see tiny black fruiting bodies in the diseased bark. Unfortunately, even with treatment, the tree is unlikely to recover, so it is best to remove it and plant a conifer that isn't susceptible to Cytospora canker. However, to try to treat Cytospora canker take the following steps:

- Prune and remove all infected branches.
- Because infection occurs through wounding, avoid injuring bark and stems.

- If pruning is necessary, do it only during dry periods.
- For a new conifer, maintain the plant's good health and vigor by ensuring good air circulation and proper care.

Phomopsis spruce decline is a recently discovered disease that appears to be caused by *Diaporthe* spp. (syn. *Phomopsis* spp.). It causes branch dieback and cankers on spruce trees, particularly Colorado spruce (*P. pungens*). Phomopsis canker is also a significant issue for drought-stressed Douglas-fir. "Flare out" of young Douglas-fir with canker disease(s) is common in parts of the Northwest, especially in south Puget Sound in Washington and in the Willamette Valley in Oregon.

Douglas-fir branches turning brown or red and dying or "flaring out" from a combination of drought stress and Phomopsis stem canker fungi. CREDIT: GLEN AHRENS

Affected trees experience tip blight (browning and dieback of branch tips), needle drop, and branch death, which usually progress over a two- to four-year period. Evidence of a canker is a distinct brown discoloration at the transition between the dying branch and healthy wood.

To treat Phomopsis spruce decline take the following steps:

- Prune off the affected area by cutting several inches below the cankered area.
- Prune in dry weather, preferably when the tree is still dormant, as spores can enter new wound cuts in spring. Unfortunately, the wilting and browning of new growth usually is not apparent until spring.
- Avoid wounding the tree.
- Grow the trees under optimum conditions.

Rusts

The two most damaging rusts in North America are **white pine blister rusts** of five-needle pines and **fusiform rust** of southern pines. White pine blister rust caused by *Cronartium ribicola* is particularly severe in the northeast, but it also causes significant mortality in the West on nonresistant white pines. Fusiform rust (*Cronartium quercuum* f. sp. *fusiforme*) is pathogenic on loblolly, slash, and longleaf pines in the southeast. To manage rust disease, select clones with genetic resistance, avoid planting in high-risk areas where alternate hosts (such as currents or gooseberries) are present, remove alternate hosts, and prune off cankers.

Root Diseases

Conifer root diseases are caused by a variety of virulent fungi that lead to root rots, such as **Armillaria root rot** (*Armillaria* spp.), **annosus root rot** (*Heterobasidion* spp.), and **laminated root rot** (*Phellinus sulphurascens*, formerly *P. weirii*). These fungi are primarily endemic components of forest ecosystems. Once established, many can persist in roots and dead trees for decades. Root rots decay the large structural roots, destroying the supporting root structure and base of a conifer, slowing the tree's growth, and eventually killing the tree. Trees infected by root rots are susceptible to windthrow (being blown down by the wind) and thus pose a hazard to humans and their dwellings, particularly in urban environments. Infected trees also become predisposed to attack by secondary fungi, bark beetles, wood borers, and other pathogens. Aboveground symptoms include thinning of the crown, excessive cone production, and resin flow at the stem or trunk. Affected conifers may also exhibit branch dieback, chlorosis (yellowing) or other foliage discoloring, and stunted growth in new foliage. However, these symptoms could be the result of environmental factors or a variety of other pests or pathogens. Closely examine the tree roots and base of the tree to determine whether the tree has root disease. For example, the roots may be brown, soft, and mushy rather than white and firm. Or, in an Armillaria root rot, at the base of the tree, you might see mushrooms or small, 1- to 2-mm-diameter, dark, string-like structures, called *rhizomorphs*, growing through the soil.

In contrast, *Phytophthora* spp. (fungus-like organisms or *oomycetes*) attack feeder roots. Symptoms of **Phytophthora root rot** include wilting, stunted foliage, chlorosis, and black, dead, feeder roots.

For help identifying the type of root disease you might be dealing with, contact your County Extension agent. For specific information on key identification features of root diseases, see Bibliography and Resources. Conifers vary in their susceptibility to a given root pathogen. Some of the most common conifer root diseases and their major hosts are shown in the table on page 193.

Armillaria root rot showing excessive resin flow at the base of the tree and white fan-shaped mycelia (fungal hyphae) under the bark in the root crown. Inset: Closeup of fungal hyphae.
CREDIT: GLEN AHRENS

Table 8.1: Root Disease/Major Hosts	
Armillaria root rot	True fir, Douglas-fir, hemlock, pine, spruce
Annosus root rot	Hemlock, true fir, pine, spruce
Laminated root rot	Douglas-fir, true fir, mountain hemlock
Black stain root disease	Douglas-fir, ponderosa pine, Jeffrey pine, piñón pine, single-leaf piñón
Phytophthora root rot	Port Orford-cedar, true fir, Douglas-fir, true cedar, arborvitae, larch, pine, spruce, yew

All five of the conifer diseases shown in the table occur in western North America, although Armillaria, with its broad range of hosts, occurs in most of North America and many other parts of the world. There is no cure for these root diseases, so the only remedy is prevention and tree removal, especially if the trees endanger people, other animals, and structures.

Strategies to prevent root disease in your conifers include the following:

- Plant conifers that are resistant to key pests in your region and compatible with your microclimate, exposure, and soil conditions.
- Plant conifers in well-draining soil. If you have heavy clay soil, be sure to mix enough compost into the entire planting area to ensure good drainage. Wet, heavy soil favors Phytophthora root rot.
- Avoid compacting the soil with heavy foot or vehicle traffic, and keep the roots cool in hot weather to help prevent Phytophthora infection. Adding 3 inches of coarse mulch or wood chips over the tree roots up to the drip line will minimize compaction and moderate soil temperatures.
- Minimize plant stress by providing good drainage and adequate light exposure, water, and space.
- Don't overwater or overfertilize conifers.
- Minimize unnecessarily injuring plants, for example, when mowing the lawn or pruning.
- Avoid moving soil from root rot areas to other sites in your garden. Remove all susceptible conifers from areas known to contain root diseases, as these diseases can spread by root contact. Note that root disease can also be spread by water, mud, and dirt transported by vehicles or foot traffic from infected areas, including forests and parks. Shoes worn in infected areas should be cleaned and disinfected before entering the garden. This applies to Phytophthora, too.

- Plant conifers resistant to root diseases. For example, Port Orford-cedar (*Chamaecyparis lawsoniana*), which is widely used in ornamental horticulture, is highly susceptible to infection by *Phytophthora lateralis*. Fortunately, highly resistant strains of *Chamaecyparis lawsoniana* are available in nurseries, including susceptible cultivars grafted onto resistant rootstock. Conifers resistant to laminated root rot, such as western white pine or western-red-cedar, are also available.

Dwarf Mistletoes

Dwarf mistletoes are parasitic flowering plants that grow exclusively on the branches and stems of conifer species in the western United States. Dwarf mistletoes infect only members of the pine family (Pinaceae) in North America, but in Europe and Asia they also infect junipers. Most commercially important conifers in western North America are susceptible to dwarf mistletoe. True mistletoes (genus *Phoradendron*), with a few exceptions, grow on hardwood trees, not conifers. Severe dwarf mistletoe infection deprives hosts of water, mineral nutrients, and often carbohydrates, reducing tree growth, wood quality, and seed and cone production, and infection will increase the host's susceptibility to pathogenic fungi and insect attacks. On the other hand, dwarf mistletoes play a critical functional role in our forests and urban ecosystems, providing excellent cover, nesting sites, and food for many birds, insects, and small mammals.

Often, dwarf mistletoes initially have an almost coral-like appearance, with short branches bearing sticky-coated, bicolored fruit. When the fruits on the female plants reach maturity, they explosively eject their seeds at about 50 miles per hour. The seeds travel 30 feet or more until they hit another conifer. Rain washes the seeds down to the base of the conifer, where they get fertilized and overwinter, infecting the host the following spring. The new dwarf mistletoe develops root-like structures called *sinkers* within the inner bark of the host, through which it gets water and nutrients, especially sucrose. Eventually, it causes overgrowth of the wood to form galls. After several years, the dwarf mistletoe produces aerial shoots, which mature to develop flowers and produce seed.

Dwarf mistletoe-infected trees are easy to spot because, after three to five years, the tree develops a witch's broom at the infection site. Witch's brooms are deformities—an unnatural, dense profusion of shoots from the infection point that resembles a broom. Not all witch's brooms are caused by dwarf mistletoes, but a telltale stem-swelling at the broom site, along with the mistletoes' aerial shoots, are sure signs that the conifer has a dwarf mistletoe infection.

Our warming climate, along with fire-suppression practices, has resulted in increasingly frequent and more extreme forest fires. Trees with numerous and large witch's brooms, as typically occur on Douglas-firs, can significantly exacerbate the fire potential in infected forest stands. Thus, it is important to manage dwarf mistletoe infections

Lodgepole pine infected with dwarf mistletoe.
CREDIT: TOM HINCKLEY

while preserving the biodiversity they facilitate, through the silvicultural practices of controlling the establishment, growth, composition, and health of forests and woodlands.

The good news is that dwarf mistletoe parasitism is relatively easy to control, particularly in our urban forests, parks, and home landscapes. Since dwarf mistletoes spread slowly, have long life cycles, and are primarily host-specific, the infections are readily detectable by the presence of stem swellings and witch's brooms. Once you observe them, immediately remove the infected branches.

Concluding Remarks

Our urban gardens give many of us an oasis in which to relax and enjoy the wonders of nature and the satisfaction and rewards of gardening. They are also essential links to the broader urban ecosystem. Unfortunately, our urban gardens are under attack by development, changes in land use, invasive plant species, and climate change. These attacks substantially change the distribution and composition of plant and animal species. Even conifers, which often tolerate adverse conditions, are harmed by the warming earth, more extreme weather, and invasive species. Now more than ever, we need to create, maintain, and conserve sustainable urban gardens. Sustainable gardens flourish over time without excessive expenditure of resources, costs, or maintenance. They can make a big difference in our urban communities because they

- provide long-term ecosystem benefits
- facilitate clean air and water
- sequester carbon to regulate the climate
- conserve energy and water
- enhance wildlife habitats
- nurture our physical and mental well-being
- bring families and communities together.

All these benefits are essential for maintaining our health. Sustainable urban gardens are those that are resilient enough to adapt to our changing environment. They help us balance nature and our human needs, now and in the future, by maintaining natural biophysical processes necessary to support plant and animal life—while promoting economic vitality.

Long-lived, drought-tolerant, adaptable conifers are a key component of sustainable landscapes, supplying many key services to the urban environment, including sequestering carbon, soaking up rainfall to minimize runoff, and removing air and water pollutants year-round. Planting trees alone, however, is not a substitute for reducing fossil fuel emissions. Each of us has a responsibility to help mitigate climate change through strategies of conservation and sustainable landscape best-practices.

Given the ecological importance of urban conifers, it is vital that each of us does our part to restore site-appropriate conifers to our urban communities. Start today. Make and execute a plan to plant a conifer (or several!) in your community. I hope you will share my passion for these exceptional plants as you build your own long-lasting, sustainable garden.

Appendix

Table 1: Conifer Cultivars Golden/Yellow in Winter and Green in Summer

Cultivar	Growth Rate [in/yr]	Form	Hardiness
Abies concolor 'Wintergold'	3–6	pyramidal	4–8
Abies homolepsis 'Winter Gold'	4–6	pyramidal	5–8
Abies koreana 'Kobalt'	4–6	pyramidal	5–8
Abies nordmanniana 'Golden Spreader'	3–5	broad cone	5–8
Pinus contorta var. *latifolia* 'Chief Joseph'	3–6	columnar	5–9
Pinus mugo 'Amber Gold'	3–4	broad mound	2–8
Pinus mugo 'Carstens'	2–4	broad mound	2–8
Pinus mugo 'Starkl'	1–2	globose	2–8
Pinus mugo 'Winter Gold'	2–4	broad mound	2–8
Pinus mugo 'Winter Sonne '	3–6	broad upright	2–8
Pinus pungens 'Johnny's Goldstrike'	4–6	pyramidal	5–9
Pinus sylvestris 'Meffengold'	5–7	pyramidal	4–8
Pinus sylvestris 'Trollguld'	2–4	broad cone	4–8
Pinus sylvestris 'Moseri'	2–4	pyramid	4–8

Table 2: Conifer Cultivars Golden/Yellow Year-Round

Cultivar	Growth Rate [in/yr]	Form	Hardiness
Abies koreana 'Aurea'	3–6	pyramidal	5–8
Abies pinsapo 'Aurea'	3–6	broad upright	6–8
Calocedrus decurrens 'Berrima Gold'	3–5	pyramidal	6–8
Chamaecyparis obtusa 'Ceramic'	2–5	pyramidal	5–8
Chamaecyparis obtusa 'Meroke'	2–5	pyramidal	5–8
Chamaecyparis obtusa 'Nana Lutea'	3–5	broad upright	5–8
Chamaecyparis pisifera 'Golden Mop'	3–6	globose	5–8
Chamaecyparis pisifera 'Harvard Gold'	3–6	globose	4–9
Cupressus macrocarpa 'Wilma Goldcrest'	7–10	narrow columnar	7–10
Juniperus horizontalis 'Mother Lode'	3–5	prostrate	3–9
Picea abies 'Amber Flickers'	2–4	pyramidal	3–8
Picea omorika 'Peve Tijn'	3–6	pyramidal	4–9
Picea orientalis 'Firefly'	3–6	pyramidal	4–8
Picea orientalis 'Skylands'	6–8	pyramidal	4–8
Picea orientalis 'Tom Thumb Gold'	1–3	globose	4–8
Picea pungens 'Sunshine'	4–6	pyramidal	3–8
Picea sitchensis 'Bentham's Sunlight'	3–6	pyramidal	5–9
Pinus parviflora 'Goldylocks'	6–10	broad upright	5–9
Pinus strobus 'Louie'	9–12	pyramidal	3–8
Pinus thunbergii 'Ogon'	9–12	pyramidal	5–9
Taxus baccata 'Standishii'	4–6	narrow columnar	6–8
Taxus cuspidata 'Nana Aurescens'	3–6	broad mount	4–8
Thuja orientalis 'Franky Boy'	3–6	pyramidal	5–8
Thuja plicata '4Ever Goldy'	10–12	columnar	5–8
Thuja plicata 'Stoneham Gold'	3–5	pyramidal	5–8
Tsuga canadensis 'Everitt Golden'	3–6	pyramidal	4–8

Table 3: Conifers of North America

Genus	Species	Common Name
Northwest Region		
Abies	*A. amabilis*	Pacific silver fir
	A. concolor	white fir
	A. grandis	grand fir
	A. lasiocarpa	subalpine fir
	A. magnifica	red fir
	A. procera	noble fir
Calocedrus	*C. decurrens*	California incense-cedar
Chamaecyparis	*C. lawsoniana*	Port Orford-cedar, Lawson-cypress
Cupressus	*C. nootkatensis*	Nooka-cypress
Juniperus	*J. communis*	common juniper
	J. occidentalis	western juniper
	J. scopulorum	Rocky Mountain juniper
Larix	*L. lyallii*	alpine larch
	L. occidentalis	western larch
Picea	*P. breweriana*	brewer spruce
	P. engelmannii	Engelmann spruce
	P. glauca	white spruce
	P. pungens	Colorado spruce
	P. sitchensis	sitka spruce
Pinus	*P. albicaulis*	whitebark pine
	P. attenuata	knobcone pine
	P. contorta var. *contorta*	shore pine
	Pinus contorta var. *Latifolia*	lodgepole pine
	P. flexilis	limber pine
	P. jeffreyi	Jeffrey pine
	P. lambertiana	sugar pine
	P. monticola	western white pine
	P. ponderosa	ponderosa pine
Pseudotsuga	*P. menziesii*	Douglas-fir
Taxus	*T. brevifolia*	Pacific yew
Thuja	*T. plicata*	western red-cedar
Tsuga	*T. heterophylla*	western hemlock
	T. mertensiana	mountain hemlock

Table 3: Conifers of North America

Genus	Species	Common Name
Southwest Region		
Abies	*A. concolor*	white fir
	A. lasiocarpa var. *arizonica*	Arizona corkbark fir
Calocedrus	*C. decurrens*	California incense-cedar
Cupressus	*C. arizonica*	Arizona cypress
	var. *glabra*	smooth Arizona cypress
	C. macrocarpa	Monterey cypress
Juniperus	*J. occidentalis*	western juniper
	J. scopulorum	Rocky Mountain juniper
Picea	*P. pungens*	Colorado spruce
Pinus	*P. albicaulis*	whitebark pine
	P. aristata	Rocky Mountain bristlecone pine
	P. attenuata	knobcone pine
	P. coulteri	coulter pine
	P. edulis	piñón pine
	P. flexilis	limber pine
	P. jeffreyi	Jeffrey pine
	P. lambertiana	sugar pine
	P. longaeva	Great Basin bristlecone pine
	P. monophylla	single-leaf pine
	P. monticola	western white pine
	P. ponderosa	ponderosa pine
Pseudotsuga	*P. macrocarpa*	bigcone Douglas-fir
	P. menziesii	Douglas-fir
Sequoia	*S. sempervirens*	coast redwood
Sequoiadendron	*S. giganteum*	giant sequoia
Torreya	*T. californica*	California torreya
Northcentral Region		
Abies	*A. balsamea*	balsam fir
Juniperus	*J. communis*	common juniper
	J. scopulorum	Rocky Mountain juniper
	J. virginiana	eastern red-cedar
Larix	*L. laricina*	tamarack, American larch
Picea	*P. glauca*	white spruce

Table 3: Conifers of North America

Genus	Species	Common Name
	P. mariana	black spruce
Pinus	*P. banksiana*	Jack pine
	P. ponderosa	ponderosa pine
	P. resinosa	red pine
	P. strobus	eastern white pine
Thuja	*T. occidentalis*	eastern arborvitae
Tsuga	*T. canadensis*	eastern hemlock
Southcentral Region		
Juniperus	*J. scopulorum*	Rocky Mountain juniper
Pinus	*P. echinata*	shortleaf pine
	P. palustris	longleaf pine
	P. ponderosa	ponderosa pine
	P. taeda	loblolly pine
Taxodium	*T. distichum*	bald-cypress
Northeast Region		
Abies	*A. balsamea*	balsam fir
Juniperus	*J. communis*	common juniper
	J. virginiana	eastern red-cedar
Larix	*L. laricina*	tamarack, American larch
Picea	*P. glauca*	white spruce
	P. mariana	black spruce
	P. rubens	red spruce
Pinus	*P. banksiana*	Jack pine
	P. pungens	Table Mountain pine
	P. resinosa	red pine
	P. rigida	pitch pine
	P. strobus	eastern white pine
Taxus	*T. canadensis*	Canadian yew
Thuja	*T. occidentalis*	eastern arborvitae
Tsuga	*T. canadensis*	eastern hemlock
	T. caroliniana	Carolina hemlock
Southeast Region		
Abies	*A. fraseri*	Fraser fir
Chamaecyparis	*C. thyoides*	Atlantic white-cypress, swamp-cedar

Table 3: Conifers of North America		
Genus	**Species**	**Common Name**
Pinus	*P. clausa*	sand pine
	P. echinata	shortleaf pine
	P. glabra	spruce pine
	P. palustris	longleaf pine
	P. pungens	Table Mountain pine
	P. rigida	pitch pine
	P. *strobus*	eastern white pine
	P. taeda	loblolly pine
Taxodium	*T. distichum*	bald-cypress
	var. *imbricatum*	pond-cypress
Torreya	*T. taxifolia*	Florida nutmeg-yew
Tsuga	*T. canadensis*	eastern hemlock

Bibliography and Resources

Albers, John J. *The Northwest Garden Manifesto: Create, Restore, and Maintain a Sustainable Yard*. Seattle, WA: Skipstone, 2018.

American Conifer Society, www.conifersociety.org.

Beutler, Linda. *The Plant Lover's Guide to Clematis*. Portland, OR: Timber Press, 2016.

Brun, Charles A., and Paula Dinius. *Selecting Plants for Screens and Hedges*. Washington State University Extension, EMO89E, 2015. https://research.wsulibs.wsu.edu/xmlui/bitstream/handle/2376/5283/EM089E.pdf?sequence=2.

Clapp, J. Casey, et al. "Rationale for the Increased Use of Conifers as Functional Green Infrastructure: A Literature Review and Synthesis." *Arboricultural Journal: The International Journal of Urban Forestry* 36 (July 2014).

Cox, Tom, and John M. Ruter. *Landscaping with Conifers and Ginkgo for the Southeast*. Gainesville, FL: University Press of Florida, 2013.

Cranshaw, Whitney, and David Shetlar. *Garden Insects of North America: The Ultimate Guide to Backyard Bugs*. 2nd ed. Princeton and Oxford: Princeton University Press, 2018.

Dirr, Michael A., and Charles W. Heuser, Jr. *The Reference Manual of Woody Plant Propagation: From Seed to Tissue Culture*. 2nd ed. Portland, OR: Timber Press, 2006.

Eckenwalder, James E. *Conifers of the World: The Complete Reference*. Portland, OR, and London: Timber Press, 2009.

Edmonds, Robert L., James K. Agee, and Robert I. Gara. *Forest Health and Protection*. 2nd ed., Long Grove, IL: Waveland Press, Inc., 2011.

Environmental Protection Agency. "Heat Island Impacts." https://www.epa.gov/heat-islands/heat-island-impacts.

Farjon, Aljos. *A Handbook of the World's Conifers*. 2 vols. Leiden, Netherlands: Koninklijke Brill NV, 2017.

Fincham, Robert L. *Small Conifers for Small Gardens*. Eatonville, WA: Coenosium Press, 2011.

Garner, R. J. *The Grafter's Handbook*. Orion Publishing Group Ltd., 1989.

Grimshaw, John, and Ross Bayton. *New Trees: Recent Introductions to Cultivation*. Royal Botanic Gardens, Kew: Kew Publishing, 2009.

Gymnosperm Database. https://www.conifers.org.

Hansen, Everett M., Kathy J. Lewis, and Gary A. Chastagner, eds. *Compendium of Conifer Diseases.* 2nd ed. St. Paul, MN: APS (American Phytopathological Society) Press, 2018.

Humphrey, Brian. *The Bench Grafter's Handbook: Principles and Practice.* 1st edition. Boca Raton, FL: CRC Press, 2019.

Johnson, Warren T., and Howard H. Lyon. *Insects that Feed on Trees and Shrubs.* Ithaca, NY: Cornell University Press, 1991.

Keane, Marc Peter. *Japanese Garden Notes: A Visual Guide to Elements and Design.* Berkeley, CA: Stone Bridge Press, 2017.

Larson, Richard A. "Grafting: A Review of Basics as Well as Special Problems Associated with Conifer Grafting." *Combined Proceedings International Plant Propagation Society* 56 (2006).

Lippke, Bruce, et al. "Life Cycle Impacts of Forest Management and Wood Utilization on Carbon Mitigation Knows and Unknowns." *Carbon Management* 22(3), (2011), 303–333.

Bruce Lippke, Maureen Puettmann, and Elaine Oneil. "Effective Uses of Forest-Derived Products to Reduce Carbon Emissions." (Updated LCI Data, December 2019), Consortium for Research on Renewable Industrial Materials (CORRIM), https://corrim.org/use-of-forest-products-to-reduce-carbon-emissions/.

Missouri Botanical Garden. "Phytophthora Root Rot of Conifers." www.missouribotanicalgarden.org/gardens-gardening/help-for-the-home-gardener/advice-tips-resources/peats-and-problems/diseases/rot/phytophthora-root-rot-of-conifers.aspx.

Peters, Emily B., and Joseph P. McFadden. "Influence of Seasonality and Vegetation Type on Suburban Microclimates." *Urban Ecosystems* 13, no. 4 (April 2010).

Rogerson Clematis Collection. www.rogersonclematiscollection.org.

Sinclair, Wayne A., and Howard H. Lyon. *Diseases of Trees and Shrubs.* Ithaca, NY: Cornell University Press, 2005.

Shaw, David C., Paul T. Oester, and Gregory M. Filip. *Managing Insects and Diseases of Oregon Conifers.* Oregon State University Extension, EM8980, June 2009. https:/catalog.extension.oregonstate.edu/em8980.

Thomas, R. William, Susan F. Mating, and Kim Tripp. *Growing Conifers: Four Season Plants.* Brooklyn, NY: Brooklyn Botanic Garden, 1997.

United States Department of Agriculture Forest Service Agriculture Handbook 727. *The Woody Plant Seed Manual.* Caldwell, NJ: Blackburn Press, 2008.

Waring, Richard H., and Jerry F. Franklin. "Evergreen Coniferous Forests of the Pacific Northwest." *Science* 204, no. 4400 (June 29, 1979): 1380–6.

Weaver, Larry, and Jenna Webster. "An Ecological Perspective for the Connoisseur Conifer Collector." *Conifer Quarterly* (Summer 2010).

World Health Organization. *Health and Sustainable Development.* https://www.who.int/sustainable-development/cities/health-risks/about/en/.

Index

Page numbers in **bold** indicate photographs, page numbers in *italics* indicate tables.

About the Authors

John J. Albers has over 30 years of experience growing and caring for conifers in a variety of garden situations. He is an educator for the Washington State Nursery and Landscape Association (WSNLA), an ecoPRO-certified sustainable landscape professional, and a former Washington State University / Kitsap County Master Gardener. He created Albers Vista Gardens, a 4-acre property containing 1,200 different species and cultivars, including 450 different conifers representing 130 different conifer species. John is author of *The Northwest Garden Manifesto* and a frequent presenter to garden clubs, master gardeners, members of the nursery and landscape industry, and the general public. He lives in Bremerton, Washington.

David E. Perry is a professional photographer who has photographed for many books, including *The Northwest Garden Manifesto* and *The 50 Mile Bouquet*, numerous magazines, *Fortune 500* corporate annual reports, and national ad campaigns for more than 35 years. He is a willing teacher and captivating storyteller whose playful, sometimes irreverent manner of speaking keeps audiences on the edge of their seats. David lives in Seattle, Washington.

ABOUT NEW SOCIETY PUBLISHERS

New Society Publishers is an activist, solutions-oriented publisher focused on publishing books for a world of change. Our books offer tips, tools, and insights from leading experts in sustainable building, homesteading, climate change, environment, conscientious commerce, renewable energy, and more—positive solutions for troubled times.

We're proud to hold to the highest environmental and social standards of any publisher in North America. When you buy New Society books, you are part of the solution!

DON'T EAT THIS BOOK (but you could)

- We print all our books in North America, never overseas
- All our books are printed on 100% **post-consumer recycled paper**, processed chlorine-free, with low-VOC vegetable-based inks (since 2002)
- Our corporate structure is an innovative employee shareholder agreement, so we're one-third employee-owned (since 2015)
- We're carbon-neutral (since 2006)
- We're certified as a B Corporation (since 2016)
- We're Signatories to the UN's Sustainable Development Goals (SDG) Publishers Compact (2020–2030, the Decade of Action)

At New Society Publishers, we care deeply about *what* we publish—but also about *how* we do business.

To download our full catalog, please visit: newsociety.com/pages/nsp-catalogue.

Sign up for New Society Publishers' newsletter for information on upcoming titles, special offers, and author events: (https://signup.e2ma.net/signup/1425175/42152/).

ENVIRONMENTAL BENEFITS STATEMENT

New Society Publishers saved the following resources by printing the pages of this book on chlorine free paper made with 100% post-consumer waste.

TREES	WATER	ENERGY	SOLID WASTE	GREENHOUSE GASES
66 FULLY GROWN	5,300 GALLONS	28 MILLION BTUs	230 POUNDS	28,700 POUNDS

Environmental impact estimates were made using the Environmental Paper Network Paper Calculator 4.0. For more information visit www.papercalculator.org.

Certified B Corporation

FSC MIX Paper from responsible sources FSC® C016245 www.fsc.org

new society PUBLISHERS www.newsociety.com